S. HRG. 115–183

THE UNITED STATES' INCREASING DEPENDENCE ON FOREIGN SOURCES OF MINERALS AND OPPORTUNITIES TO REBUILD AND IMPROVE THE SUPPLY CHAIN IN THE UNITED STATES

HEARING

BEFORE THE

COMMITTEE ON ENERGY AND NATURAL RESOURCES UNITED STATES SENATE

ONE HUNDRED FIFTEENTH CONGRESS

FIRST SESSION

MARCH 28, 2017

Printed for the use of the
Committee on Energy and Natural Resources

Available via the World Wide Web: http://www.govinfo.gov

U.S. GOVERNMENT PUBLISHING OFFICE
24–976   WASHINGTON : 2018

For sale by the Superintendent of Documents, U.S. Government Publishing Office
Internet: bookstore.gpo.gov   Phone: toll free (866) 512–1800; DC area (202) 512–1800
Fax: (202) 512–2104   Mail: Stop IDCC, Washington, DC 20402–0001

COMMITTEE ON ENERGY AND NATURAL RESOURCES

LISA MURKOWSKI, Alaska, *Chairman*

JOHN BARRASSO, Wyoming	MARIA CANTWELL, Washington
JAMES E. RISCH, Idaho	RON WYDEN, Oregon
MIKE LEE, Utah	BERNARD SANDERS, Vermont
JEFF FLAKE, Arizona	DEBBIE STABENOW, Michigan
STEVE DAINES, Montana	AL FRANKEN, Minnesota
CORY GARDNER, Colorado	JOE MANCHIN III, West Virginia
LAMAR ALEXANDER, Tennessee	MARTIN HEINRICH, New Mexico
JOHN HOEVEN, North Dakota	MAZIE K. HIRONO, Hawaii
BILL CASSIDY, Louisiana	ANGUS S. KING, JR., Maine
ROB PORTMAN, Ohio	TAMMY DUCKWORTH, Illinois
LUTHER STRANGE, Alabama	CATHERINE CORTEZ MASTO, Nevada

COLIN HAYES, *Staff Director*
PATRICK J. MCCORMICK III, *Chief Counsel*
ANNIE HOEFLER, *Professional Staff Member*
SEVERIN WIGGENHORN, *Senior Counsel*
ANGELA BECKER-DIPPMANN, *Democratic Staff Director*
SAM E. FOWLER, *Democratic Chief Counsel*
SPENCER GRAY, *Democratic Professional Staff Member*
MELANIE STANSBURY, *Democratic Professional Staff Member*

CONTENTS

OPENING STATEMENTS

	Page
Murkowski, Hon. Lisa, Chairman and a U.S. Senator from Alaska	1
Cortez Masto, Hon. Catherine, a U.S. Senator from Nevada	3

WITNESSES

Hitzman, Dr. Murray, Associate Director–Energy and Minerals, U.S. Geological Survey, U.S. Department of the Interior	4
Barrios, Alf, Chief Executive, Rio Tinto Aluminum	12
Hinde, Dr. Chris, Director, Reports, S&P Global Market Intelligence	21
MacGillivray, Randy, Vice President Project Development, Ucore Rare Metals, Inc.	25
Cosgriff, Vice Admiral Kevin J., USN (Retired), President and CEO, National Electrical Manufacturers Association	30
Eggert, Dr. Roderick G., Viola Vestal Coulter Foundation Chair in Mineral Economics, Division of Economics and Business, Colorado School of Mines	36

ALPHABETICAL LISTING AND APPENDIX MATERIAL SUBMITTED

Barrios, Alf:
    Opening Statement 12
    Written Testimony 14
    Responses to Questions for the Record 74
Cortez Masto, Hon. Catherine:
    Opening Statement 3
Cosgriff, Vice Admiral Kevin J.:
    Opening Statement 30
    Written Testimony 32
    Responses to Questions for the Record 86
Eggert, Dr. Roderick G.:
    Opening Statement 36
    Written Testimony 38
    Responses to Questions for the Record 88
Hinde, Dr. Chris:
    Opening Statement 21
    Written Testimony 23
    Responses to Questions for the Record 76
Hitzman, Dr. Murray:
    Opening Statement 4
    Written Testimony 6
    Responses to Questions for the Record 68
MacGillivray, Randy:
    Opening Statement 25
    Written Testimony 27
    Responses to Questions for the Record 81
(The) Minerals Science and Information Coalition:
    Statement for the Record 94
Murkowski, Hon. Lisa:
    Opening Statement 1

THE UNITED STATES' INCREASING DEPENDENCE ON FOREIGN SOURCES OF MINERALS AND OPPORTUNITIES TO REBUILD AND IMPROVE THE SUPPLY CHAIN IN THE UNITED STATES

TUESDAY, MARCH 28, 2017

U.S. SENATE,
COMMITTEE ON ENERGY AND NATURAL RESOURCES,
Washington, DC.

The Committee met, pursuant to notice, at 10:00 a.m. in Room SD–366, Dirksen Senate Office Building, Hon. Lisa Murkowski, Chairman of the Committee, presiding.

OPENING STATEMENT OF HON. LISA MURKOWSKI, U.S. SENATOR FROM ALASKA

CHAIRMAN. Good morning. The Committee will come to order.

I understand Senator Cantwell will be coming later, but Senator Cortez Masto will be subbing in this morning. We appreciate that a great deal.

We are here today to receive testimony on the United States' foreign mineral dependence. It will probably come as no surprise to anyone here that, I believe, this is a significant and a growing threat to our nation. Resolving it and restoring our mineral security is a priority for me and many members of this Committee.

Our starting point is to recognize that minerals are important because they are the building blocks of our modern society, from the smallest computer chips to the tallest skyscrapers and just about everything in-between.

Minerals are fundamental to fracking, MRI machines, and jet engines. The homes that we live in, the food we eat, the cars we drive, and the computers we use, all depend on minerals. Almost every product in our nation is made from, or uses, minerals, yet more and more these minerals are now being produced somewhere else.

According to the U.S. Geological Survey (USGS), we imported at least 50 percent of our supply of 50 different minerals, including 100 percent of our supply of 20 of them, just last year in 2016. That is a major increase from our foreign dependence levels in 1978 when this data was first collected, and it suggests that we are on the verge of replacing our dependence on foreign oil with an equally, if not even more damaging, dependence on foreign minerals.

Rare earth elements are perhaps the best-known example. With the Mountain Pass Mine in California now closed, we once again import 100 percent of our supply of rare earths, exposing us to potential supply shortages and price volatility while reducing our international leverage and attractiveness for manufacturing. It is the same story with graphite, palladium, indium, manganese, niobium and many others.

When you look at the list of what it is that we import, where we import it from, and what it is used for, it quickly becomes clear that we have a problem on our hands. Our foreign mineral dependence is a threat to our ability to create jobs in this country. It limits our growth, our competitiveness, and our national security. It may seem abstract right now for some who are not responsible for managing a supply chain, but there will come a day when it will become real for all of us when we simply cannot acquire a mineral or when the market for a mineral changes so dramatically that entire industries are affected.

Some agencies have begun to wake up to the threats posed by our foreign mineral dependence, but on the whole, the Federal Government is not paying anywhere near enough attention. Executive agencies are not as focused or as coordinated as they need to be, and they do not have the direction or authority that they need to make lasting progress to restore our mineral security.

That is why, for the past three Congresses, I have introduced legislation to improve our nation's mineral security along with members from both sides of our Committee. Last Congress we included our work in our broad, bipartisan energy bill which both the Committee and the Senate overwhelmingly approved.

As we examine policy options in this new Congress, I remain convinced that our ideas on minerals are on the right track and they are as timely as ever.

I continue to believe that we should have a mechanism to track which minerals are critical in use and susceptible to supply disruption. When a mineral is listed as critical, we should survey our lands to determine the extent of our resource base.

When it comes to permitting delays for new mines, our nation is among the worst in the world, so fixing our broken system is one of the single most important steps we can take.

We should also promote research into alternatives, efficiency, and recycling options, especially for minerals that we do not have in significant abundance.

We should build out our minerals forecasting capability to provide a better understanding of mineral-related trends and early warnings when problems do arise.

And we need to pay attention to workforce issues so that smart kids are taught by qualified professionals and can go on to find success in environmentally-responsible mining operations.

This Congress offers a perfect opportunity to finally bring our minerals policies into the 21st century and to begin to restore our nation's mineral security. Today we start that effort by focusing on the importance of minerals, the threats posed by our rising foreign dependence, and a discussion of the solutions that are within our reach. So I look forward to hearing from each member of the panel this morning.

I will now turn to Senator Cortez Masto and welcome her for her comments.

STATEMENT OF HON. CATHERINE CORTEZ MASTO, U.S. SENATOR FROM NEVADA

Senator CORTEZ MASTO. Thank you.

I want to thank Madam Chair Murkowski and Ranking Member Cantwell for bringing together this hearing that is incredibly salient to our interest in not only rebuilding our mining industry but in retooling our economy for a high-tech world.

The United States is at a critical stage of innovation. The technologies that we all use demand a steady supply of critical minerals, minerals that are primarily imported with an increasing global demand.

When our dependence on foreign minerals increases and we are 100 percent import reliant for 20 minerals, including 8 identified as critical, it is absolutely necessary to prioritize the security of our supply chain.

We have the opportunity, right now, to seize on mineral supply independence as we have in the energy sector with fuel. Our country has the supplies, workforce, technology and government programs to rebuild our domestic supply, but they require investment.

Not only does improving our supplies ensure our mining industry's success, but it will also improve our economy, other important industries and resilience to global competition.

Mining companies provide thousands of good jobs for residents in Nevada, pay millions of dollars in tax revenues and help support other parts of our state's economy.

Additionally, mining companies like Barrick and Newmont not only employ thousands of Nevadans but also prioritize digital improvements that increase efficiency, transparency and corporate sustainability. The ripple effect of an expanded domestic mining industry includes technology companies, research institutions, energy systems and the military.

Technology minerals are absolutely critical for many of the technologies that are part of our everyday lives and stand to improve our energy systems from our cell phones, to solar panels and battery storage. Leveraging our resources is a real opportunity which, if done responsibly, continues the charge of my state and the country into a great age of innovation and resiliency in a competitive global market.

But know that there are challenges that we must address. I am eager to hear from our esteemed experts who will inform us about the challenges they face or the solutions they believe will move us forward. I know that investments in technologies, research, education and a trained workforce and improving the permitting review process, all are priorities moving forward as our country increases its domestic supply of critical minerals and the innovation dependent upon those resources.

Thank you very much for joining us today.

CHAIRMAN. Thank you, Senator.

We will now turn to our witnesses. Thank you. I appreciate not only your input this morning but what you have done in the various sectors and spaces that you operate. Your leadership is greatly appreciated.

We are going to start off this morning with Dr. Murray Hitzman, who is the Associate Director for Energy and Minerals at the U.S. Geological Survey. Welcome to you, Mr. Hitzman.

He will be followed by Mr. Alf Barrios, who is the Chief Executive of Rio Tinto Aluminum. Welcome.

Dr. Chris Hinde is the Director of Reports, Metals and Mining at S&P Global Market Intelligence. We thank him for being here.

Next is a friend of mine, Mr. Randy MacGillivray, who is the Vice President of Project Development at Ucore Rare Metals, Incorporated. Welcome.

We are joined by Vice Admiral Kevin Cosgriff, U.S. Navy Retired. He is the President and CEO of the National Electric Manufacturers Association (NEMA). We appreciate you being here.

Rounding out the panel is Dr. Roderick Eggert, who is the Viola Vestal Coulter Foundation Chair in Mineral Economics at the Division of Economics and Business at the Colorado School of Mines. We appreciate your contributions this morning.

Dr. Hitzman, we will ask you to lead off the panel. I would ask each of you to limit your comments to five minutes. Your full testimony will be incorporated as part of the record, and we will hold our questions until each of you has spoken. I look forward to your input this morning.

Dr. Hitzman, welcome.

STATEMENT OF DR. MURRAY HITZMAN, ASSOCIATE DIRECTOR–ENERGY AND MINERALS, U.S. GEOLOGICAL SURVEY, U.S. DEPARTMENT OF THE INTERIOR

Dr. HITZMAN. Good morning, Chairman Murkowski, and thanks to the members of the Committee. Thank you for the opportunity to be here to testify about the nation's foreign mineral dependence.

The U.S. Geological Survey is responsible for conducting research and collecting data on a wide variety of mineral resources. The USGS collects, analyzes and disseminates information on current production and consumption of 84 mineral commodities, both domestically and internationally for 180 countries. These data include information on domestic production and use, import sources, world production capacity, and recycling. These mineral data are published annually in the Mineral Commodities Summaries.

Global demand for mineral commodities is on the rise, and the United States is increasingly reliant on foreign sources for raw processed mineral materials. In 2016, our studies show that imports made up more than one-half of the U.S. apparent consumption of 50 non-fuel mineral commodities valued at $32.3 billion. The United States was 100 percent reliant for 20 of these mineral commodities, including 8 identified as critical. This is an increase from 2015 when the country was more than 50 percent dependent on 47 non-fuel mineral commodities and 100 percent reliant on 19.

The list of mineral commodities for which the United States is 100 percent import reliant includes both well-known and obscure commodities. Elements that the U.S. depends on from foreign sources include the rare-earth element, Europium, which is essential for getting a bright red color out of a TV screen and metal oxides that are responsible for some popular automobile paint colors.

The metal oxides are an example of the effect of supply disruptions. For several months after the 2011 Japanese earthquake and tsunami, American vehicle manufacturers were unable to supply customers with popular red and black sports cars and trucks due to the unavailability of a critical mineral ingredient.

In 2015, the USGS, in cooperation with the Department of Energy, developed a screening tool to identify critical minerals of concern for economic and national security and to stay ahead of technology changes and geopolitical unrest. This criticality tool accounts for several variables in identifying critical minerals, including how vulnerable the supply chain is to disruption, how much production growth is expected for the material and market dynamics. These studies allow the users to rank minerals from lower to higher potential criticality. The resultant rankings are being used today by the Defense Logistics Agency.

An accurate assessment of the nation's mineral resources must include not only the resources available in the ground but also those that become available through recycling. Metal supply consists of primary material from a mining operation and secondary material which is composed of new and old scrap. Although recycling is a significant source of some non-fuel mineral resources such as aluminum, technical difficulties with recycling mean that for other mineral commodities such as the rare earths, recycling is extremely challenging.

In addition to providing information on mineral production and consumption, the USGS also produces data that aids in assessing the mineral potential of the nation. For example, the USGS recently released a study on critical minerals in Alaska.

To help source minerals domestically, the USGS undertakes both geologic mapping and the production of regional-scale geophysical maps that help define areas favorable for mineral exploration.

Currently only about one-third of the United States has been mapped at the detailed scales required for mineral exploration. Other countries, such as Canada and Australia, have undertaken such geological and geophysical surveys and have reported that investments of $1 by the government have resulted in further investments of over $5 by the private sector.

The Department, through the USGS, stands ready to fulfill its role as the federal provider of unbiased research on known mineral resources, assessment of undiscovered mineral resources, data to aid mineral exploration by the private sector and information on domestic and global production and consumption of mineral resources for use in global critical mineral supply chain analysis.

Thank you for the opportunity to testify. I'm very happy to answer any questions.

[The prepared statement of Dr. Hitzman follows:]

Statement of Murray Hitzman
Associate Director – Energy and Minerals, U.S. Geological Survey
U.S. Department of the Interior
before the
Senate Energy and Natural Resources Committee
on
March 28, 2017

Good morning Chairman Murkowski, Ranking Member Cantwell, and Members of the Committee, and thank you for the opportunity to discuss the Nation's foreign mineral dependence.

Background

The U.S. Geological Survey (USGS) is responsible for conducting research and collecting data on a wide variety of mineral resources. Research is conducted to understand the geologic processes that have concentrated known mineral resources at specific localities in the Earth's crust and to assess quantities, qualities, and areas of undiscovered mineral resources, or potential future supply. USGS mineral commodity specialists collect, analyze, and disseminate data and information that document current production and consumption for 84 mineral commodities, both domestically and internationally for 180 countries. These data include information on domestic production and use, import sources, world production capacity, and recycling. The data allow for a comprehensive understanding of the complete life cycle of mineral resources and materials. These mineral data are published annually in the *Mineral Commodities Summaries*. The most recent installment for 2017 was released in January.

Global demand for mineral commodities continues to be on the rise. Mineral commodities have ever more applications in consumer and national security products especially those involving advanced technologies. The United States remains a major mineral producer with an estimated total value of non-fuel mineral resources of $75.6 billion and is net exporter of 16 non-fuel mineral commodities. However the country also is increasingly reliant on foreign sources for raw processed mineral materials. In 2016, imports made up more than one-half of the U.S. apparent consumption of 50 non-fuel mineral commodities (valued at $32.3 billion), and the United States was 100% import reliant for 20 of these mineral commodities (valued at $1.3 billion), including 8 identified as critical minerals. This is an increase from 47 non-fuel mineral commodities on which the country was more than one-half dependent in 2015 and 19 non-fuel commodities for which the country was 100% reliant in 2015. China, followed by Canada, supplied the largest number of non-fuel mineral commodities to the U.S. in 2016, similar to the case in 2015.

2016 U.S. NET IMPORT RELIANCE

Commodity	Percent	Major import sources (2012–15)[2]
ARSENIC	100	China, Japan
ASBESTOS	100	Brazil
CESIUM	100	Canada
FLUORSPAR	100	Mexico, China, South Africa, Mongolia
GALLIUM	100	China, Germany, United Kingdom, Ukraine
GRAPHITE (natural)	100	China, Mexico, Canada, Brazil
INDIUM	100	Canada, China, France, Belgium
MANGANESE	100	South Africa, Gabon, Australia, Georgia
MICA, sheet (natural)	100	China, Brazil, Belgium, Austria
NIOBIUM (columbium)	100	Brazil, Canada
QUARTZ CRYSTAL (industrial)	100	China, Japan, Romania, United Kingdom
RARE EARTHS	100	China, Estonia, France, Japan
RUBIDIUM	100	Canada
SCANDIUM	100	China
STRONTIUM	100	Mexico, Germany, China
TANTALUM	100	China, Kazakhstan, Germany, Thailand
THALLIUM	100	Germany, Russia
THORIUM	100	India, France, United Kingdom
VANADIUM	100	Czech Republic, Canada, Republic of Korea, Austria
YTTRIUM	100	China, Estonia, Japan, Germany
GEMSTONES	99	Israel, India, Belgium, South Africa
BISMUTH	95	China, Belgium, Peru, United Kingdom
TITANIUM MINERAL CONCENTRATES	91	South Africa, Australia, Canada, Mozambique
POTASH	90	Canada, Russia, Chile, Israel
GERMANIUM	85	China, Belgium, Russia, Canada
STONE (dimension)	84	China, Brazil, Italy, Turkey
ANTIMONY	83	China, Thailand, Bolivia, Belgium
ZINC	82	Canada, Mexico, Peru, Australia
RHENIUM	81	Chile, Poland, Germany
GARNET (industrial)	79	Australia, India, South Africa, China
BARITE	78	China, India, Morocco, Mexico
FUSED ALUMINUM OXIDE (crude)	>75	China, Canada, Venezuela
BAUXITE	>75	Jamaica, Brazil, Guinea, Guyana
TELLURIUM	>75	Canada, China, Belgium, Philippines
TIN	75	Peru, Indonesia, Malaysia, Bolivia
COBALT	74	China, Norway, Finland, Japan
DIAMOND (dust grit, and powder)	73	China, Ireland, Romania, Russia
PLATINUM	73	South Africa, Germany, United Kingdom, Italy
IRON OXIDE PIGMENTS (natural)	>70	Cyprus, France, Austria, Spain
IRON OXIDE PIGMENTS (synthetic)	>70	China, Germany, Canada, Brazil
PEAT	69	Canada
SILVER	67	Mexico, Canada, Peru, Poland
CHROMIUM	58	South Africa, Kazakhstan, Russia
MAGNESIUM COMPOUNDS	53	China, Brazil, Canada, Australia
ALUMINUM	52	Canada, Russia, United Arab Emirates, China
IODINE	>50	Chile, Japan
LITHIUM	>50	Chile, Argentina, China
SILICON CARBIDE (crude)	>50	China, South Africa, Netherlands, Romania
ZIRCONIUM MINERAL CONCENTRATES	>50	South Africa, Australia, Senegal
ZIRCONIUM (unwrought)	>50	China, Japan, Germany
BROMINE	<50	Israel, China, Jordan
MICA, scrap and flake (natural)	49	Canada, China, India, Finland
PALLADIUM	48	South Africa, Russia, Italy, United Kingdom
TITANIUM (sponge)	41	Japan, Kazakhstan, China
SILICON	38	Russia, China, Canada, Brazil, South Africa
COPPER	34	Chile, Canada, Mexico
LEAD	30	Canada, Mexico, Republic of Korea, Peru
VERMICULITE	30	Brazil, South Africa, China, Zimbabwe
MAGNESIUM METAL	<30	Israel, Canada, China, Mexico
NITROGEN (fixed)—AMMONIA	28	Trinidad and Tobago, Canada, Russia, Ukraine
TUNGSTEN	>25	China, Canada, Bolivia, Germany
NICKEL	25	Canada, Australia, Norway, Russia

**MAJOR IMPORT SOURCES OF NONFUEL MINERAL COMMODITIES
FOR WHICH THE UNITED STATES WAS GREATER THAN 50% NET IMPORT RELIANT IN 2016**

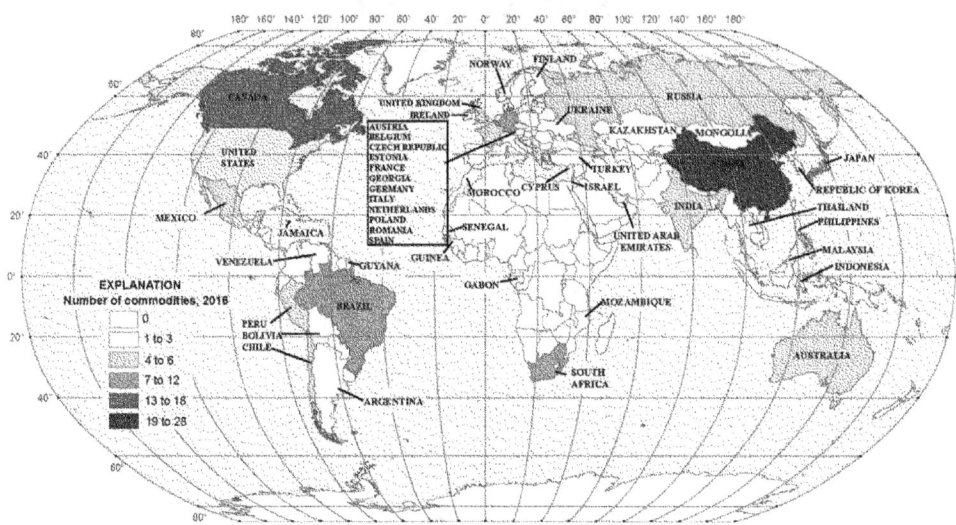

The list of non-fuel mineral commodities for which the United States is 100% import reliant includes some well-known commodities such as manganese and rare earth elements as well as some more obscure commodities such as gallium and niobium. The rare earth elements are currently produced almost exclusively in China though domestic sources do exist, including the recently reopened and then shuttered Mountain Pass, California mine.

The USGS continues to research the occurrence of rare earth element deposits in the United States (a 2010 USGS study documented 28 rare earth deposits in the United States that potentially could be developed) and explore geological processes that may form domestic deposits that are yet to be discovered. For example, USGS scientists are conducting research in the southeastern United States on granites that contain high concentrations of rare earth elements to understand and assess likely chemical and physical processes that could lead to the enrichment and retention of rare earth elements in soil and to characterize the minerals in which heavy rare earth elements reside in regolith. The project will develop criteria and methodologies to delineate the occurrence of rare earth element-clay resources and define characteristics that relate to sustainable mining of rare earth element clay deposits. In addition, the USGS recently released a study on critical minerals, including rare earth elements, in Alaska.

The element gallium is recovered as a byproduct of processing bauxite (the material from which aluminum is extracted) and zinc ores primarily in China (80% of worldwide low-grade gallium capacity). Gallium is used primarily to manufacture gallium-arsenide wafers used in integrated circuits for defense applications and high-performance computers, light emitting diodes (LEDs), and solar cells.

Other exotic elements that the U.S. depends on from foreign sources include europium, which is essential for getting a bright red color out of TV screens and metal oxides responsible for some popular automobile paint colors. As an example of the effect of supply disruptions, after the 2011 Japanese earthquake and tsunami, for several months American vehicle manufacturers were unable to supply customers with popular red and black sports cars and trucks due to the unavailability of a critical ingredient.

In 2015, mineral specialists in the USGS National Minerals Information Center, with cooperation from the Department of Energy, developed an early warning screening tool to identify critical minerals of concern for economic and national security and stay ahead of the curve as technology changes and geopolitical unrest shifts.[1] The tool accounts for several variables in identifying critical minerals, including how vulnerable the supply chain is to disruption, how much production growth is expected for the material, and market dynamics. Once the system has filtered out minerals that are not "potentially critical," the remaining minerals receive further analysis. In-depth studies allow users to rank each mineral from lower to higher potential criticality. The resultant rankings are currently used by the Defense Logistics Agency (DLA) to define a cutoff point for analyzing potentially critical materials for shortfalls. Tom Rasmussen, the Director of Strategic Plans for the DLA, has stated that "The USGS is world renowned as having an incredible reputation for providing mineral information. Having the USGS brand name on this early warning system lends [it] a great deal of credibility."

An accurate assessment of the Nation's mineral resources must include not only the resources available in the ground but also those that become available through recycling. Metal supply consists of primary material from a mining operation and secondary material, which is composed of new and old scrap. Recycling can contribute to metal production. Metals show a wide range of recycling rates, recycling efficiency, and new-to-old-scrap ratios. Recycling rates cluster in the range from 15 to 45 percent for different resources. Although recycling is a major source of some non-fuel mineral resources such as aluminum, technical difficulties with recycling mean that for other mineral commodities such as the rare earth elements recycling is challenging. USGS compiles information about recycling but research on new methods of metal recycling is undertaken mainly by the Department of Energy.

In addition to providing information on mineral production and consumption, the USGS also produces data that aids in assessing the mineral potential of the country, which we have done since 1879. This work continues as different mineral commodities gain importance for the economy and as our understanding improves of how mineral deposits form and how they can be discovered. Geological maps are a primary source of information for mineral exploration. Many USGS geological maps are produced in conjunction with state geological surveys through the National Cooperative Geologic Mapping Program through cooperative agreements.

---

[1] The tool was featured in a report to Congress submitted in 2016 by the Interagency Subcommittee on Critical and Strategic Mineral Supply Chains and entitled, *Assessment of Critical Minerals: Screening Methodology and Initial Application*.

To help source minerals domestically, the USGS undertakes both geologic mapping and the production of regional scale geophysical maps such as aeromagnetic and radiometric maps that help define areas favorable for exploration. This work generally requires more detailed geologic mapping, and currently about one-third of the United States has been mapped at these scales. Other countries such as Canada and Australia have undertaken such geological and geophysical surveys nationwide and have reported that investments of one dollar by the government have resulted in further investment of over five dollars by the private sector.

Conclusion

The Department maintains a workforce of geoscientists, including geologists, geochemists, geophysicists, and resource specialists, with expertise in critical minerals and materials. The Department continuously collects, analyzes, and disseminates data and information on domestic and global rare earth and other critical mineral reserves and resources, production, consumption, and use. This information is published annually in the USGS *Mineral Commodity Summaries* (USGS, 2017) and includes a description of current events, trends, and issues related to supply and demand. These data inform analyses and policies concerning the Nation's dependence on foreign sources of mineral commodities.

The Department, through the USGS, stands ready to fulfill its role as the federal provider of unbiased research on known mineral resources, assessment of undiscovered mineral resources, and information on domestic and global production and consumption of mineral resources for use in global critical mineral supply chain analysis.

Thank you for the opportunity to present on behalf of the Department on the important subject of mineral resources. I will be happy to answer any questions.

For More Information

Duke, J.M., 2010, Government geoscience to support mineral exploration: public policy rationale and impact: Prospectors and Developers Association of Canada. Toronto, Canada, 64 p.

Foley, N. and Ayuso, R., 2015, REE enrichment in granite-derived regolith deposits of the Southeastern United States: Prospective source rocks and accumulation processes, In: Simandl, G.J. and Neetz, M., (Eds.), Symposium on Strategic and Critical Materials Proceedings, November 13-14, 2015, Victoria, British Columbia, British Columbia Ministry of Energy and Mines, British Columbia Geological Survey Paper 2015-3, pp. 131-138.
http://www.empr.gov.bc.ca/Mining/Geoscience/PublicationsCatalogue/Papers/Documents/P2015-3/16%20Foley.pdf

Goonan, T.G., 2011, Rare Earth Elements—End Use and Recyclability: U.S. Geological Survey SIR 2011-5094 available at http://pubs.usgs.gov/sir/2011/5094/

Goonan, T.G., 2012, Lithium use in batteries: U.S. Geological Survey Circular 1371, 14 p., 2011–1042, 11 p., available at http://pubs.usgs.gov/circ/1371/

Goonan, T.G., 2012, Materials flow of indium in the United States in 2008 and 2009: U.S. Geological Survey Circular 1377, 12 p., available at http://pubs.usgs.gov/circ/1377/

Long, K.R., Van Gosen, B.S., Foley, N.K., and Cordier, Daniel, 2010, The principal rare earth elements deposits of the United States—A summary of domestic deposits and a global perspective: U.S. Geological Survey Scientific Investigations Report 2010–5220, 96 p., available at http://pubs.usgs.gov/sir/2010/5220/

Menzie, W.D., Baker, M.S., Bleiwas, D.I., and Kuo, Chin, 2011, Mines and mineral processing facilities in the vicinity of the March 11, 2011, earthquake in northern Honshu, Japan: U.S. Geological Survey Open-File Report 2011–1069, 7 p. (http://pubs.usgs.gov/of/2011/1069/.)

National Research Council, 2008, Minerals, Critical Minerals, and the U.S. Economy: Washington, D.C., National Academies Press, 264 p.

Soto-Viruet, Yadira, Menzie, W.D., Papp, J.F., and Yager, T.R., 2013, An exploration in mineral supply chain mapping using tantalum as an example: U.S. Geological Survey Open-File Report 2013–1239, 51 p., http://pubs.usgs.gov/of/2013/1239/

Tse, Pui-Kwan, 2011, China's Rare-Earth Industry. U.S. Geological Survey Open-File Report 2011–1042, 11 p., available only at http://pubs.usgs.gov/of/2011/1042.

USGS, 2017, Mineral Commodity Summaries 2016. U.S. Geological Survey, 202 p. https://minerals.usgs.gov/minerals/pubs/mcs/2017/mcs2017.pdf

Wilburn, D.R., 2012, Byproduct metals and rare-earth elements used in the production of light-emitting diodes—Overview of principal sources of supply and material requirements for selected markets: U.S. Geological Survey Scientific Investigations Report 2012–5215, 15 p., available online at http://pubs.usgs.gov/sir/2012/5215/.

Yager, T.R., Soto-Viruet, Yadira, and Barry, J.J., 2012, Recent strikes in South Africa's platinum-group metal mines—Effects upon world platinum-group metal supplies: USGS Open-File Report 2012-1273, 18 p., http://pubs.usgs.gov/of/2012/1273/.

CHAIRMAN. Thank you, Dr. Hitzman.

I believe that other members of the Committee have received the USGS report on the Alaska assessment which I found was very helpful with the maps, thank you.

Mr. Barrios, welcome.

STATEMENT OF ALF BARRIOS, CHIEF EXECUTIVE, RIO TINTO ALUMINUM

Mr. BARRIOS. Good morning, Chairman Murkowski and members of the Committee.

My name is Alf Barrios. I am the Chief Executive of Rio Tinto Aluminum. I sit on the Rio Tinto Executive Committee and serve as the company's country sponsor for Canada and the United States.

Rio Tinto has been operating in the U.S. for over 100 years. Our operations include Kennecott Copper in Salt Lake City, Utah; Resolution Copper in Superior, Arizona; and Rio Tinto Boron in California.

The most recent Mineral Commodity Summaries by the U.S. Geological Survey should set off alarm bells in the White House and Congress. The study, published earlier this year, indicates the U.S. is now import-dependent for 50 different metals and minerals and 100 percent import-dependent for 20. The trend is troubling.

U.S. mineral dependency is at a record-high, now double what it was 20 years ago. During that same timeframe, investment in minerals exploration projects has dropped from 20 percent to seven percent. This drift away from greater self-sufficiency for the basic building blocks of our economy compromises our economic and national security and ignores North America's rich reserves of metals and minerals that are at the front-end of the manufacturing supply chain. Dependence on imported essential materials to meet the needs of key domestic industries leaves the U.S. unnecessarily vulnerable to disruptions to vital supply chains.

Of course, no country, not even the United States, is blessed with top tier deposits of every essential mineral. Enhancing the U.S.' ability to access its own resources does not mean we should raise barriers to imported materials. Nowhere are the mutual benefits of trade more apparent than the integrated supply chains in North America where imports from Canada make U.S. manufacturers more competitive and vice versa.

We have a real opportunity to realize the full potential of the domestic mining industry. Clearly demands for minerals is increasing as global population expands and minerals are used in a greater range of applications, particularly associated with the deployment of new technologies.

The manufacturing sector has expressed heightened concerns about securing access to the minerals they need when they need them. According to a survey of 400 manufacturing executives, more than 90 percent are concerned about supply disruptions, citing geopolitics and increasing global demand as the most pressing factors. In addition, 80 percent of U.S. manufacturing leaders recognize the importance of sourcing domestic minerals and metals, noting strengthened national security as reasons for doing so.

An outdated, inefficient permitting system presents a major barrier to the domestic mining sector's ability to perform to its full potential and supply more of our infrastructure needs. The U.S. has one of the longest permitting processes in the world for mining projects. In the U.S., necessary government authorizations now take approximately seven to ten years to secure, placing the U.S. at a competitive disadvantage in attracting investment for mineral development. By comparison, permitting in Australia and Canada, which have similar environmental standards and practices as the U.S., takes between two and three years.

Authorities, ranging from the National Academy of Sciences to the Departments of Energy and Defense to international mining consulting firms, have identified permitting delays as among the most significant risk and impediments to mining projects in the United States. Most recently, the U.S. Government Accountability Office linked the need to streamline the mine permitting process to mitigate supply risks.

To address supply chain vulnerability and import dependence, President Trump and Congress should continue to examine ways to improve permitting of new U.S. mines and smelters. The mining industry strongly supports efforts in the House and Senate to address the mine permitting process including S. 145, the National Strategic and Critical Minerals Production Act. The bill provides for efficient, timely, and thorough permit reviews and incorporates best practices for coordination between state and federal agencies.

We also appreciate the efforts by Chairman Murkowski last Congress to move forward the American Mineral Security Act. Her legislation, cosponsored by many on this Committee, was a step forward in bringing the U.S. in line with its global peers who are preparing to meet the 21st century challenges of mineral supply chain reliability and security.

I would like to conclude by reemphasizing the important role the mining industry has in supporting U.S. manufacturing and infrastructure development, but also by acknowledging that Rio Tinto understands responsibility extends far beyond.

We must lead by example when it comes to community engagement, reclamation and pioneering technology innovation. For example, on Lake Chelan in north-central Washington State, we have been working to rehabilitate the old copper mine which we obtained through a large acquisition in 2008. Despite never commercially benefiting from the mine, Rio Tinto has brought its global expertise to the project and has spent hundreds of millions of dollars to rehabilitate Holden Village.

Thank you for the opportunity to testify today. I appreciate the Committee's leadership on this very important issue.

[The prepared statement of Mr. Barrios follows:]

Testimony of

Alf Barrios
Chief executive, Rio Tinto Aluminum
before the
United States Senate Committee on Energy and Natural Resources

Hearing on
*The United States' Increasing Dependence on Foreign Sources of
Minerals and Opportunities to Rebuild and Improve the Supply Chain
in the United States*

March 28, 2017

Good morning Chairman Murkowski and members of the Committee.

My name is Alf Barrios and I am the Chief executive of Rio Tinto Aluminum. I sit on the Rio Tinto Executive Committee and serve as the company's country sponsor for Canada and the United States.

Rio Tinto is a global mining company operating in 35 countries with 50,000 employees and we are particularly proud to have been operating in the United States for over 100 years.

As the Rio Tinto U.S. country sponsor, I oversee and coordinate all of our activities across the U.S., including our operations: Kennecott Copper in Salt Lake City, Utah, Resolution Copper in Superior, Arizona and Rio Tinto Boron in California.

I appreciate the opportunity to provide this testimony on a very serious, yet often ignored issue, the United States' increasing dependence on foreign sources of minerals. I am pleased that the hearing goes beyond a discussion of the problem to seek solutions for rebuilding and improving supply chains in the United States.

Growing Mineral Import Reliance is a Troubling Trend

The most recent *Mineral Commodity Summaries* by the U.S. Geological Survey (USGS) should set off alarm bells in the White House and Congress. The study, published earlier this year, indicates that the United States is now import-dependent for 50 different metals and minerals – and 100 percent import-dependent for 20.[1] That's half of the naturally-occurring elements on the Periodic Table. The trend line is troubling: U.S. mineral dependency is at a record-high, now double what it was 20 years ago. During that same timeframe, investment in minerals exploration projects has dropped from 20 percent to 7 percent. This drift away from greater self-sufficiency for the basic building blocks of our economy compromises our economic and national security and ignores North America's rich reserves of metals and minerals that are the front-end of the manufacturing supply chain.

Dependence on imported essential materials to meet the needs of key domestic industries, such as manufacturing, leaves the United States unnecessarily vulnerable to disruptions to vital supply chains. Today, U.S.

---

[1] USGS, *Mineral Commodity Summaries 2017*, available at https://minerals.usgs.gov/minerals/pubs/mcs/2017/mcs2017.pdf

manufacturers rely on imported minerals to meet more than half their needs. If key minerals or metals are suddenly unavailable – due to political instability in a source country, shipping disruptions or restrictions on mining access – the whole supply chain could grind to a halt.

These trends are unsustainable in a highly competitive world economy in which the demand for minerals continues to increase and stability of supply is a growing concern. This point was underscored by KPMG in a report that looked at sustainability megaforces and predicted by 2030 that 83 billion tons of minerals, metals and biomass will be extracted from the earth, 55 percent more than in 2010. The study authors conclude: "the message is clear; over the next 20 years, demand for material resources will soar while supplies will become increasingly difficult to obtain."[2]

We have a real opportunity to realize the full potential of the domestic mining industry. Clearly demand for minerals is increasing as global population expands and minerals are used in a greater range of applications, particularly associated with the deployment of new technologies. The mining industry is poised to provide even greater contributions to the economy building upon the 2016 value added to America's gross domestic product (GDP). In 2016, the value added by major industries consuming mineral materials is $2.78 trillion – nearly 15 percent of U.S. GDP.[3]

Manufacturing and technology sectors have expressed heightened concerns about securing access to the minerals they need when they need them. According to a survey of 400 manufacturing executives, more than 90 percent are concerned about supply disruptions, citing geopolitics and increasing global demand as the most pressing factors. In addition, 80 percent of U.S. manufacturing leaders recognize the importance of sourcing domestic minerals and metals, noting decreased dependence on foreign minerals and metals and strengthened national security as reasons for doing so. Nearly 85 percent believe a strong domestic supply chain of critical minerals and metals will ensure job creation and economic growth in America.[4]

As for re-shoring American manufacturing capability, the Rand Corporation has documented the threats to U.S. manufacturing from our increasing mineral import reliance. In a 2013 study, Rand warns this situation makes "U.S. manufacturers vulnerable to export restrictions. that can result in two-

---

[2] Expect the Unexpected: Building business value in a changing world – KPMG, 2012
[3] USGS, *Mineral Commodity Summaries 2017*
[4] Edelman Berland, Survey of U.S. Manufacturing Executives (September 2014).

tier pricing, under which domestic manufacturers in the producing country have access to materials at lower prices than those charged for exports, thereby hindering the international competitiveness of U.S. manufacturers and creating pressure to move manufacturing away from the U.S. and into the producing country.[5]

Permitting Delays Are the Most Significant Impediment to Providing Additional Domestic Supplies of Minerals for Infrastructure Projects

An outdated, inefficient permitting system presents a major barrier to the domestic mining sector's ability to perform to its full potential and supply more of our infrastructure needs. The U.S. has one of the longest permitting processes in the world for mining projects. In the U.S., necessary government authorizations now take approximately seven to 10 years to secure, placing the U.S. at a competitive disadvantage in attracting investment for mineral development. By comparison, permitting in Australia and Canada, which have similar environmental standards and practices as the U.S., takes between two and three years.

Authorities ranging from the National Academy of Sciences to the Departments of Energy and Defense to international mining consulting firms have identified permitting delays as among the most significant risks and impediments to mining projects in the United States.[6] Most recently, the U.S Government Accountability Office linked the need to streamline the mine permitting process to mitigate supply risks.[7]

These delays have real consequences. The National Mining Association (NMA) commissioned a study that will be discussed in more detail by the witness from S&P Global Market Intelligence that demonstrates empirically the destruction of value which results from unnecessary, extended delays to project development.[8]

While not included in the S&P study, Rio Tinto's Resolution Copper project is currently in the permitting process. This world class copper deposit represents one of the largest undeveloped copper resources in the world and is anticipated to have a 50-year mine life that will support thousands of

---

[5] Rand Corporation. *Critical Materials: Present Danger to U.S. Manufacturing,* (p. ix), 2013
[6] *See* National Resources Council, *Hardrock Mining on Federal Lands,* National Academy Press (1999); U.S. Department of Energy, *Critical Materials Strategy* (Dec. 2010); U.S. Geological Survey USGS, *the Principal Rare Earth Elements Deposits of the United States—A Summary of Domestic Deposits and a Global Perspective,* 2010; Behre Dolbear, *Where Not to Invest* (2015).
[7] GAO Report 16-699, *Advanced Technologies: Strengthened Federal Approach Needed to Help Identify and Mitigate Supply Risks for Critical Raw Materials,* Dec. 2016
[8] SNL Metals & Mining, *Permitting, Economic Value and Mining in the United States,* June 2015.

jobs annually and many millions in tax revenues. The U.S. Forest Service is the lead regulator for the project and has been a constructive and responsive partner in the NEPA review process. The NEPA review process was started in November of 2013. While Rio Tinto has spent over $1.3 billion on the Resolution Project for permitting, studies and project shaping, the project is years away from a final permit. In other countries, a project entering review in late 2013 would be in the last laps of the permitting process – or even commencing production.

Solutions are Necessary

The efficiency and predictability of the permitting process matters in decisions about where companies chose to invest. Adverse public policies such as the U.S. Environmental Protection Agency's proposal to duplicate state and federal financial assurance programs can also be significant deterrents to investment and the development of a sustainable resource sector.

To address supply chain vulnerability and import dependence, President Trump and Congress should continue to examine ways to improve the permitting of new U.S. mines and smelters, eliminate duplicative regulations and support policies that encourage resource and materials innovation.

There is strong public support for policies that enable the use of domestic resources for infrastructure. In fact, a new poll conducted this week reveals that 71 percent of voters support using domestically-sourced materials for infrastructure and that 65 percent support enacting policies such as shorter permitting timeframes for mining projects to ensure timely access to important minerals and metals that build steady and stable supply chains.[9] Manufacturing executives are equally supportive of ensuring efficient permitting as 95 percent of executives surveyed are worried that the lag in the permitting process for new mines has a serious impact on their competitiveness.[10]

Legislative action has an important role to play as well. The mining industry strongly supports efforts in the House and Senate to address the mine permitting process, including S. 145, the National Strategic and Critical

---

[9] Polling Shows Strong Support for Policies that Encourage the Use of American Minerals in U.S. Infrastructure, Manufacturing
http://nma.org/2017/03/20/polling-shows-strong-support-for-policies-that-encourage-the-use-of-american-minerals-in-u-s-infrastructure-manufacturing/
[10] Edelman Berland, Survey of U.S. Manufacturing Executives (September 2014).

Minerals Production Act, which offers proactive solutions to fix the U.S. permitting process. The legislation carefully addresses the deficiencies of the outdated U.S. permitting system without changing environmental and other protections afforded by current laws and regulations. The bill provides for efficient, timely and thorough permit reviews and incorporates best practices for coordination between state and federal agencies.

We also appreciated the efforts by Chairman Murkowski last Congress to move forward the American Security Minerals Act. Her legislation, cosponsored by many on this committee, was a step forward in bringing the US in line with its global peers who are preparing to meet the 21st century challenges of mineral supply chain reliability and security.

I would like to conclude by reemphasizing the important role the mining industry has in supporting US manufacturing and infrastructure development, but to also acknowledge that Rio Tinto understands responsibility extends far beyond. We must lead by example when it comes to community engagement, reclamation and pioneering technology innovation.

For example, on Lake Chelan in north-central Washington State we have been working to rehabilitate the old Holden copper mine, which we obtained through a larger acquisition in 2008. Despite never commercially benefiting from the mine, Rio Tinto has brought its global expertise to the project and has spent hundreds of millions of dollars to rehabilitate Holden Village, which now serves as a spiritual retreat and community center.

We also have our eye towards the future and we are pursuing ways of improving America's mineral footprint to boost resource innovation. At Rio Tinto's Garfield copper smelter in Utah, we are partnering with the U.S. Department of Energy's Critical Materials Institute (CMI) to find new ways to fully recover and recycle the minerals that future technologies will require. This means not just looking at more efficient ways to process and extract minerals from the ground, but also "urban mining" of electronic waste. One of the most concentrated sources of valuable metals is in old phones and electronics sent off to scrap. To address this waste of resources, we are testing technology that could help capture the valuable minerals in electronic waste in the copper smelting process – including copper's critical and strategic co-products, like rhenium and even rare earths, materials used in alternative energy and fighter aircraft, in smart phones and smart bombs. While clearly accelerating demand cannot be fully met by increasing recycling or substitution, we believe recycling is an important component of our corporate sustainability efforts.

Thank you for the opportunity to testify today. I appreciate the committee's leadership on this very important issue.

CHAIRMAN. Thank you, sir.
Let's go to Dr. Hinde. Welcome.

STATEMENT OF DR. CHRIS HINDE, DIRECTOR, REPORTS, S&P GLOBAL MARKET INTELLIGENCE

Dr. HINDE. Chair and Members of the Committee, good morning and thank you for inviting me to present to this Committee.

My written testimony can be summarized in six parts.

First, if I may, what authority can I bring to bear? I've been writing about the mining industry for 30 years at S&P Global Market Intelligence, as the world's largest database of exploration and mining activity.

Second, a comment about the supply/demand scene in the USA. S&P Global has conducted two relevant studies within the past three years. In September 2014, we argued that a healthy, local exploration and mining sector is important for the American economy. And in mid–2015, we quantified the impact of permitting delays on mine development in the USA.

The second of these reports identified the destruction of value of the results from even short delays in the permitting process whilst the first reports demonstrated a clear mismatch in the USA between consumption and the local supply of required metals and minerals.

This country is still the world's largest economy and per capita metals consumption in the USA is far in excess of the citizens of other countries. In contrast, the USA ranks as only the seventh largest mining nation by value of production. This shortfall is especially regrettable because manufacturing activity is returning to the USA. This move is driven by manufacturers' desire to reduce the risks in their supply chains and a consumer's increasing concerns regarding corporate accountability. We found that the USA miners are highly efficient and generally apply best practices with regard to productivity, sustainability, and safety.

Third, mining is a very uncertain business with geology and mineral endowment being extremely difficult to assess, and its companies being price takers rather than having the luxury of being able to set the price of their products. Because of this extra risk, the industry acquires financial returns that are in excess of most of business activities.

The fourth of my six points is the USA offers some key advantages to miners including a stable political and economic environment, but most companies with comparable mineral resources and similar environmental standards offer a much more certain permitting process. Like companies and industries the world over, mining executives simply seek certainty in the legal and fiscal processes that they face.

As one of my colleagues just mentioned, it takes, on average, seven to ten years to secure the permits needed for mines to reach production in the USA. In contrast and with very similar overall requirements, Canada and Australia are managing their average permitting periods of barely two years.

In the USA, many agencies and stakeholders are involved with a requirement for multiple permits and rather undefined goals for indigenous groups, the general public, and non-governmental organizations. Rigorous permitting is, of course, necessary and is to a similar standard to our knowledge in the USA, Canada, and Australia; however, the permitting process is much better defined in Canada and Australia with a shorter timeline for the various agencies to respond.

Fifth, a quick note on the global scene. In our corporate exploration strategies we report, we identified $7.2 billion of global, nonferrous exploration last year. That's not including iron ore and coal exploration. Only $500 million of this was spent here on exploration compared with very nearly $900 million in Australia and close to $1 billion in Canada. Indeed, the USA exploration expenditure has fallen from the record $1.7 billion spent on exploration locally in 2012. So current exploration is running at a third of the record level.

Finally, an observation from a foreigner. The USA remains highly prospective from a geological point of view. Unfortunately, the country's existing permitting system presents a formidable barrier to the development of its own mineral wealth. This has left the USA unnecessarily dependent on local mines, whose remaining life is declining or on foreign sources of metals and mineral resources.

Your country and its mining industry would benefit from a more streamlined permitting process, ideally, something similar to those already being applied by the world's leading mining nations.

Thank you.

[The prepared statement of Dr. Hinde follows:]

Dr Chris Hinde; Director, Reports; S&P Global Market Intelligence

Testimony to the Senate Energy and Natural Resources Committee
Washington; Tuesday, March 28, 2017

Increasing USA Dependence on Foreign Sources of Metals and Minerals
S&P Global Market Intelligence has the world's largest database of exploration and mining activity. In addition to this comprehensive database, which covers some 30,000 projects and 3,000 mining companies, the Metals & Mining division offers a news service and consulting business on the extractive sector.

We previously conducted an independent study to quantify the impact of permitting delays on mine development in the USA. The report demonstrated, empirically, the destruction of value that results from increases in project risk, particularly delays in the permitting process. For example, the Kensington gold mine in Alaska was plagued by permitting issues during development, and commenced production in 2010, fully 17 years after the originally planned start date. Similarly, the Rosemont copper project in Arizona has still not secured its mining permit seven years after the originally mooted commencement of production in 2010.

New mines can typically lose over one-third of their economic value as a result of even relatively small delays in reaching production. Extended delays can render the investment unviable for less robust ventures. In such circumstances, even apparently large mineral deposits can become uneconomic to mine.

Mining is a business where certainty is important as building mines is a capital-intensive process. Uncertainties do exist as geology and mineral endowment are extremely difficult to assess accurately in advance. Also, and more than for most other sectors, mining companies are subject to the vagaries of global markets – they are generally price 'takers' rather than being able to enjoy the benefit of setting prices. Because of this extra risk, the industry requires financial returns in excess of most other business sectors. As such, the industry looks to the business climate of potential investment nations to help minimize project risks.

The USA offers some key advantages to miners, such as a stable political and economic environment. However, most countries with comparable mineral resources offer a much more certain permitting process. Like companies in all industries the world over, mining executives simply seek certainty in the legal and fiscal processes that they face. Miners have enough uncertainty in their hunt for resources without the extra burden of over-complicated, or unclear, routes to development once a mineable resource has been identified.

It takes on average seven to ten years to secure the permits needed for mines to reach production in the USA. In contrast, Canada and Australia (countries with similarly rich natural resources and equally stringent environmental regulations), have average permitting periods of barely two years. In the USA, multiple agencies and stakeholders are involved, with a requirement for multiple permits and rather undefined roles for indigenous groups, the general public and nongovernmental organizations. This necessary process is much better defined in Canada and Australia, with a shorter timeline for the various agencies to respond, and the responsibility for preparing a stringent environmental review lies with the mining company, not the government.

S&P Global Market Intelligence's latest annual Corporate Exploration Strategies report identified almost 1,600 companies that had spent at least US$100,000 on exploration for minerals last year. The total amount spent by these companies in 2016 fell 21% from the previous year to US$6.9 billion, which is barely one-third of the amount spent on exploration in 2012 after four years of falling expenditure. In the USA, exploration investment dropped by 30% in 2016. The total amount spent globally on non-ferrous exploration last year was an estimated US$7.2 billion.

Of the global total, only US$500 million was spent on exploration last year in the USA, compared with US$897 million in Australia and US$971 million in Canada (14% of the global total). The USA exploration expenditure has averaged 7% to 8% of the global total in recent years, and the amount has fallen from the record US$1.7 billion spent on exploration locally in 2012.

Our independent research has previously established why a healthy exploration and mining sector is important for the USA economy. There is a clear mismatch between the country's mineral consumption and the local supply of these metals and minerals. The USA still boasts the world's largest economy and, according to the Minerals Education Coalition, each citizen born last year is expected to consume almost 3.13 Mlb of metals and minerals over their lifetime.

Despite a decline in work on infrastructure in the USA over the past decade, this per capita consumption by today's Americans still includes nearly 21,300 lb of iron ore and 950 lb of copper over their lifetime. This is far in excess of the consumption of metals and minerals by the citizens of every other country. In contrast, the USA ranks as only the seventh largest mining nation by value of its production.

Another key finding of S&P Global Market Intelligence's earlier research was that manufacturing activity was returning to the USA, driven by manufacturers' desire to reduce the risks in their supply chains and of consumers' increasing concerns regarding corporate accountability. Consumers want to see evidence of sustainable production processes, the use of recycled materials and of sound environmental practice. We found that, relative to their global peers, USA miners are highly efficient, and generally apply best practices with regard to productivity, sustainability and safety.

The USA remains highly prospective, from a geological point of view, with abundant mineral resources that are of high quality. Unfortunately, the country's duplicative, inefficient and uncertain permitting system presents a formidable barrier to American companies' ability to deliver on their skills and access to local minerals. This has left the USA unnecessarily dependent on local mines whose remaining life is declining, or on foreign sources of metals and mineral resources.

The solution is relatively simple. The USA has abundant resources of metals and minerals, and it has the companies and people with the skill to extract these natural resources efficiently and cleanly. Rigorous permitting will always be required to ensure appropriate exploitation of a nation's wealth, and to monitor the application of best practice. What the country, and its mining industry, needs is to adopt those more streamlined permitting processes that are already being applied by the world's leading mining nations.

CHAIRMAN. Thank you, Doctor.
Mr. MacGillivray, welcome.

STATEMENT OF RANDY MACGILLIVRAY, VICE PRESIDENT PROJECT DEVELOPMENT, UCORE RARE METALS, INC.

Mr. MACGILLIVRAY. Madam Chair Murkowski, Acting Ranking Member Cortez Masto, and distinguished Members of the Committee, I would first like to thank you for the invitation to testify before you today. It's a great honor to testify before the United States Senate and I hope to provide you with some valuable information regarding the state of the industrial base for the production of strategic and critical materials in the United States from the perspective of a domestic miner.

I presently serve as the Vice President of Project Development for Ucore Rare Metals, a junior mining company with a rare-earth element project located in Southeast Alaska. Ucore is currently developing its Bokan-Dotson Ridge Rare Earth Project which presents the opportunity for near-term recovery of crucial, heavy, rare earth elements. Located in Alaska, the project would give the U.S., the world's leading consumer of rare earth elements, strategic access to a domestic supply.

The issue of foreign mineral dependence is not new, but its importance cannot be overstated. At present, the People's Republic of China dominates the production of numerous metals, including rare earth elements, which are essential for the proper function of everything from smartphones in our pockets, to advanced weapons systems used by the modern warfighter. In fact, China exhibits a near monopoly on the production of these materials introducing a dangerous risk into our supply chains.

Meanwhile, the U.S. has no operating producer of rare earth elements after the highly-publicized bankruptcy and closure of the only domestic rare earth mine in 2015. To date, the sole mitigation strategy adopted by the U.S. has been to stockpile small reserves of materials deemed to be critical and to promote substitution and recycling efforts, an inadequate approach given the criticality of these materials. Without a U.S. supply base, should the Chinese ever decide to curtail the supply of these materials to the U.S., we would be left without access, endangering both our domestic economy and our military.

Furthermore, Chinese production of these materials often relies on outdated and environmentally destructive mining and processing practices. The solvent extraction separation process used extensively by the Chinese to recover rare earths has a low selectivity for individual elements, necessitating the use of numerous separation stages using highly corrosive chemicals and generating vast amounts of toxic and radioactive waste for which very little care is taken in disposal.

To witness firsthand the toll that Chinese rare earth production is having on the environment, one need not look farther than the artificial lake located in China's Inner Mongolia region where black chemical sludge, a byproduct of solvent extraction, stains the landscape. This embrace of environmental pollution on behalf of the Chinese, in combination with the lack of worker protections, allows the Chinese to manipulate the market and effectively control global prices. Chinese producers have willingly undercut the rare-earth price driving foreign competition out of the market while the Chinese government has refused to address illegal mining and trading operations which have led to greater supply, lower prices, and further consolidation of rare earth production in China.

In light of the current situation and American dependence on these materials, the need for domestic sources and production is paramount to ensuring our national security; however, Chinese market manipulation over the past decade and notable failed domestic projects have left capital markets unwilling to fund critical material projects. Domestic mining and separation firms, with advancements in environmentally friendly technologies, would benefit from support to bridge the divide between operating on a pilot scale and full commercialization of the new technology.

The technologies to secure American independence in the critical materials markets exist, but government needs to be the key to unlocking the door for a domestic supply of critical materials for energy and defense applications.

Congress has previously been supportive of the domestic mining sector as seen by the introduction of legislation last Congress by the Madam Chair which would have promoted the development of green technology to meet the nation's demand for critical materials. Ucore remains fully committed to solving the critical materials issues facing our country and working toward solutions developed in coordination with Congress to alleviate our dependence on foreign nations for these materials.

Thank you.

[The prepared statement of Mr. MacGillivray follows:]

Statement of Randy MacGillivray
Vice President Project Development
Ucore Rare Metals, Inc.
Before the Senate Energy and Natural Resources Committee
March 28, 2017

Chairman Murkowski, Ranking Member Cantwell, and distinguished Members of the Committee, I would first like to thank you for the invitation to testify before you today. It is a great honor to testify before the United States Senate and I hope to provide you with valuable information regarding the state of the industrial base for the production of strategic and critical materials in the United States from the perspective of domestic miners.

I presently serve as the Vice President of Project Development for Ucore ("Ucore") Rare Metals, a junior mining company with a rare earth element project located in southeast Alaska. Ucore is currently developing its Bokan – Dotson Ridge Rare Earth Project which presents the opportunity for near term recovery of crucial heavy rare earth elements. Located in Alaska, the project would give the U.S., the world's leading consumer of rare earth elements, strategic access to a domestic supply.

The issue of foreign mineral dependence is not new but its importance cannot be overstated. At present, the People's Republic of China dominates the production of numerous materials, including rare earth elements, which are essential for the proper function of everything from the smart phones in our pockets to advanced weapons systems used by the modern warfighter. In fact, China exhibits a near monopoly on the production of these materials introducing a dangerous risk into our supply chains. Meanwhile, the U.S. has no operating producer of rare earth elements after the highly publicized bankruptcy and closure of the only domestic rare earth mine in 2015. To date, the sole mitigation strategy adopted by the U.S. has been to stockpile small reserves of materials deemed to be critical and to promote substitution and recycling efforts, an inadequate approach given the criticality of these materials. Without a U.S. supply base, should the Chinese ever decide to curtail the supply of these materials to the U.S. we would be left without access endangering both our domestic economy and our military.

Furthermore, Chinese production of these materials often relies on outdated and environmentally destructive mining and processing practices. The solvent extraction separation process used extensively by the Chinese to recover rare earths has a very low selectivity for individual elements necessitating the use of numerous separation stages using highly corrosive chemicals and generating vast amounts of toxic and radioactive waste for which very little care is taken in disposal. To witness firsthand the toll Chinese rare earth production is having on the environment one need not look further than the artificial lake located in China's Inner Mongolia region where black chemical sludge, a byproduct of solvent extraction, stains the landscape. This embrace of environmental pollution on behalf of the Chinese, in combination with the lack of worker protections, allows the Chinese to manipulate the market and effectively control global prices. Chinese producers have willingly undercut the prices of foreign competition driving them out of the market while the government has refused to address illegal mining and trading operations invariably leading to greater supply, lower prices, and further consolidation of production in China.

In light of the current situation and American dependence on these materials, the need for domestic sources and production is paramount to ensuring our national security. However, Chinese market manipulation over the past decade and notable failed domestic projects have left capital markets unwilling to fund critical material projects. Domestic mining and separation firms, despite advancements in environmentally friendly technologies, enabling the clean separation of critical materials, would benefit from Federal support to initiate commercialization of a new separation technology. Technologies that would secure American independence in the critical materials markets exist, but government needs to be the key to unlocking the door for a domestic supply of critical materials for energy and defense applications.

Congress has previously been supportive of the domestic mining sector as seen by the introduction of legislation last Congress by the Chairman which would have promoted the development of green technology to meet the nation's demand for critical materials. Ucore remains fully committed to solving the critical materials issues facing our country and working towards solutions developed in coordination with Congress to alleviate our dependence on foreign nations for these materials.

Ucore presents a unique opportunity to both invest in our domestic manufacturing base, which would spur job creation and support the local economy, and solve a pressing national security issue. Investment in the aforementioned Bokan – Dotson Ridge Rare Earth Project would not only provide the U.S. with a domestic supply of rare earth material but also support an estimated 190 families and deliver $18 million in annual payroll. Furthermore, Ucore has embraced the adoption of green technologies capable of separating rare earths from virgin ore without using harsh chemicals, limiting the impact on the environment.

Molecular Recognition Technology ("MRT") is a Nobel Prize winning technology that has been adapted by Ucore for use separating rare earth elements. MRT is a self-contained separation process capable of separating the entire suite of rare earth elements at greater than 99 percent purity. In partnership with IBC Advanced Technologies, Ucore invested in the development of ligands specific to rare earth chemistry and successfully incorporated MRT into its Plan of Operations for the rare earth mine project. Ucore and IBC have constructed and successfully operated a pilot scale plant to separate individual rare earths using the Bokan - Dotson Ridge ore as the feedstock material.

Since the successful completion of the pilot plant, Ucore staff have been evaluating the potential to source alternate rare earth feedstock material to supply a proof of concept commercial-demonstration scale rare earth separation plant using MRT. The ideal alternate feedstock would be sourced from by-product resources from existing mining operations. The natural attributes of MRT have allowed Ucore to identify profitable, niche opportunities to produce precious and specialty metals in very specific joint venture arrangements. These opportunities, however, given the current the Chinese controlled market, may not address national, metals security issues without federal support to scale projects focused explicitly on the critical material requirements of the U.S. military complex. Ucore is confident in the ability of MRT to meet the needs of the

U.S. government but remains cognizant that successful commercialization sometimes arrives too late to address the problem. Government support to take MRT from pilot scale production to commercialization should not be seen as proving out a new technology, but rather as facilitating the resolution of a growing problem. Given the current geo-political landscape, foreign access to rare earths could be restricted at any moment and while the technology to mitigate this problem exists, government should take the precaution to ensure that the domestic industrial base is established if such a time comes as the U.S. loses access to foreign sources of rare earths.

Government empowerment of a local supply of rare earth material, in combination with proven separation technology, MRT, would enable the U.S. to initially reduce and ultimately eliminate any dependence on foreign producer rare earths. This action is necessary to ensure that the U.S. has continued and uninterrupted access to materials essential for defense and national security purposes. Ucore is proud to be an industry leader in this effort and looks forward to working with Congress on solutions to this complex problem.

CHAIRMAN. Thank you, Mr. MacGillivray.
Vice Admiral Cosgriff, welcome.

STATEMENT OF VICE ADMIRAL KEVIN J. COSGRIFF, USN (RETIRED), PRESIDENT AND CEO, NATIONAL ELECTRICAL MANUFACTURERS ASSOCIATION

Admiral COSGRIFF. Thank you, Madam Chairman. Members of the Committee, we appreciate the opportunity for the National Electrical Manufacturers Association to appear here today on this important subject.

To put that into context, that's some 350 member companies in the electro industry, as we like to call it, and also importantly the medical imaging industry. Industry-wide that's some 400,000 American jobs with 7,000 facilities in every state of the Union. It's approximately $114 billion a year in production and $50 billion in exports.

As you might expect, NEMA supports policies that provide greater assurance to our companies of stable, continuous, and affordable inputs for their domestic manufacturing. Challenging supply conditions and price volatility in those inputs can be a significant concern to U.S. companies in multiple sectors, including our own. When we speak to our members, supply chain risk management is very much on their minds every day.

While some of our companies source raw materials, many are one or more steps away from that and purchase processed or semi-processed material that's more ready for the manufacturing effort. But one way or another, every one of our companies is dependent on reliable access to raw materials.

In the area of rare earths, the supply crisis several years ago has eased due to multiple factors, including some changes in technology and also the market that has been commented on, including in China. But U.S. firms still remain largely dependent on shipments from China for rare earths.

Foreign sourcing of lithium, not a rare earth, but nonetheless an important element, is significant as well, although not absolute.

We're also hearing from our members about the state of the U.S. aluminum industry, and factors that are leading to occasional constrained conditions. A number of our manufacturers of electrical wire and cable report that previous suppliers have either gone out of business or are otherwise operating at reduced capacity.

Copper, as you might expect, is another key metal, and about one-third of the total used is from overseas.

In the area of medical imaging there is a metal substance of essential importance, specifically Molybdenum–99, I'll call it Moly 99, and its parent is the parent isotope of Technetium–99m, call that Tech 99, is used in approximately 40,000 diagnostic procedures a day. Tech 99 has a very short half-life and therefore must be produced on a continuous basis. The U.S. consumes about half of the world's Moly 99 and has no domestic source. Canada, which used to supply the U.S. with half of our needs, ceased routine production last year.

In 2012, Congress enacted S. 99, the American Medical Isotopes Production Act as part of the Defense Authorization bill. We commend this Committee for its work on S. 99, and we encourage its oversight responsibilities to monitor implementation of this law so that patients can get the right scan at the right time.

Returning to the bigger picture, we support a federal role in minerals policy, including research and development, as well as minerals information and analysis. It's important to add that a balanced mineral policy is an important support to domestic manufacturing and employment.

Despite many efforts to date, many manufacturers' dependence on foreign sources of critical minerals, including rare earths and other raw materials, remains a concern. Companies manage this risk by diversifying supplies and, if possible, holding more inventory, both of which can impact operating costs and therefore, competitiveness. Having access to more secure, price competitive supplies closer to home, domestic as well as the other NAFTA countries, or Western Hemisphere, more broadly, is desirable.

At the end of the day, the issue we are discussing is about whether the U.S. electro and medical imaging companies can manufacture what they need to manufacture here at home. Clearly this involves access to minerals, related information, and a regulatory environment that helps them compete globally.

Thank you again for this opportunity and I look forward to any questions you may have.

[The prepared statement of Admiral Cosgriff follows:]

National Electrical Manufacturers Association

The Association of Electrical Equipment
and Medical Imaging Manufacturers
www.nema.org

Prepared Testimony of
Kevin J. Cosgriff
President and CEO
National Electrical Manufacturers Association (NEMA)

Hearing to examine the United States' increasing dependence on foreign sources
of minerals and opportunities to rebuild and improve the supply chain in the United States

Senate Committee on Energy and Natural Resources
March 28, 2017

Chairman Murkowski and Ranking Member Cantwell, and Members of the Committee,

Thank you for the invitation and opportunity to provide the following remarks on behalf of the Members of the National Electrical Manufacturers Association (NEMA) on the topic of minerals availability and the importance of minerals access for the U.S. electroindustry and medical imaging manufacturers.

NEMA represents 350 electrical and medical imaging manufacturers at the forefront of safety, reliability, and efficiency. Our combined industries account for more than 400,000 American jobs and more than 7,000 facilities across the United States, including Alaska, Washington, Nevada, and the other 47 states. Domestic production exceeds $114 billion per year and exports top $50 billion.

NEMA supports policies that provide greater assurance to our companies of stable, continuous and affordable supplies of inputs for domestic manufacturing. We also support legislation, regulations and resulting processes (e.g., permitting) that are at the minimum essential, transparent, easily understood and quickly accomplished.

Supply conditions and price volatility of basic inputs can be a significant challenge to U.S. companies in multiple sectors. When we speak with our Member companies about these issues, they are understandably somewhat reticent as competition is intense. Innovation is a real advantage and it is ongoing in our companies' laboratories, design centers, and testing facilities daily. Suffice to say supply chains are part of this competitive environment – especially for cutting-edge technology – and therefore supply chain risk management is often proprietary.

That said, while some of our Member companies source raw materials, many companies are one or more steps away from that and purchase semi-processed or processed material more ready for industry and their workers will have access to the minerals, related information and financial environment they need to be globally competitive.

Thank you again for the opportunity to provide these brief remarks and I would be pleased to consider any questions Members of the Committee may have.

development of new Mo-99 production facilities. However, industry must now convert its technology from highly enriched uranium (HEU) to non-HEU sources. While removing HEU from medical isotope production is important for non-proliferation, concern remains that the conversion has placed additional costs on the industry, patients and their providers.

NEMA thanks the Committee for its work on S. 99 and encourages the Committee to use its oversight authority to monitor implementation of AMIPA to ensure a domestic source of Molybdenum-99 that will allow patients to get the right scan at the right time.

Returning to the bigger picture, we support the federal role in critical minerals policy, including research and development as well as minerals information and analysis. It is important to add that critical minerals policy is an important contributor to domestic manufacturers and workers.

We commend the work of the research communities at the Critical Minerals Institute at Ames National Laboratory and the Colorado School of Mines. Partnerships with other institutions, including some of our companies, have helped to accelerate progress and advance the state of knowledge in many areas, including supply diversification, development of substitute materials, more-efficient use of critical materials and the challenges of reclamation and recycling. We were gratified to see CMI announcement earlier this month of a new partnership aimed at lithium-ion battery recycling.

The Minerals Commodity Summaries and other minerals information and analysis work published by the USGS Minerals Information Service is bedrock material on which NEMA and our industries' economic forecasters rely.

Let me conclude by saying that greater supply chain awareness has inspired many of our companies to institute sustainability programs. These vary by company but can include measures such as diversion from the waste stream during the manufacturing process and reutilization of pre-consumer raw materials, such as poly-vinyl chloride (PVC) and copper, as well as recycling of aluminum and steel products.

However, despite significant efforts to date, many manufacturers' dependence on foreign sources of critical minerals, including rare earths and other raw materials, remains a concern. Companies manage this risk by diversifying sources of supply, if possible, and holding more inventory, both of which can impact operating costs and overall competitiveness. Having access to more secure, price competitive supplies closer to home is desirable.

At the end of the day, the issue we are discussing today is about whether U.S. electroindustry and medical imaging companies will be able to manufacture products and where they will be able to manufacture them. Our companies are working to meet the Nation's future needs in energy, health care and transportation. NEMA would support initiatives to improve the prospects that U.S.

CHAIRMAN. Thank you very much for your testimony.
Finally, Dr. Eggert, welcome.

STATEMENT OF DR. RODERICK G. EGGERT, VIOLA VESTAL COULTER FOUNDATION CHAIR IN MINERAL ECONOMICS, DIVISION OF ECONOMICS AND BUSINESS, COLORADO SCHOOL OF MINES

Dr. EGGERT. Thank you. Good morning.

I have three key points in my testimony. First, government plays an essential role in fostering domestic supply chains of raw materials through research and education which, in turn, are important determinants of innovation. Second, both recycling and new mines will be important in meeting future raw material challenges. And third, I would suggest that it is risky imports rather than import dependence itself that is the problem, and in turn, risky imports are but one aspect of the larger issue of supply chain risks and long term resource availability.

Now consider research, education, and recycling in turn and starting with research.

Two aspects of research, I believe, are especially worthy of government involvement. First, early stage research and development which the private sector acting alone is likely to underinvest in from the perspective of society as a whole because its benefits are risky, far in the future and difficult for private companies to fully capture. Second, activities aimed at facilitating the transition, the conversion of new knowledge to commercial products and applications insights from basic research often languish because of insufficient communication between basic researchers and commercial developers of new technology.

More narrowly, and with respect to extracting and recovering materials from both mineral deposits and wastes, I believe there are two grand research challenges. The first is chemical separations. The challenge of separating one element from another in a mineral deposit or in a waste material. The second, resource efficiency, optimizing the recovery of multiple elements from the same mineral deposit or from the same waste product.

Turning to education. Part of the issue with education in this area is basic science, technology, engineering and math, but part of it is discipline specific. The dearth of resource discipline graduates in fields like economic geology, mineral processing, extractive metallurgy and even material science and engineering is highlighted by a 2013 National Research Council study.

With respect to recycling for the major metals, iron and steel, aluminum, copper, lead, zinc, there are already well-established recycling industries and recycling already plays an important role in the supply chain.

For minor metals, however, very little recycling occurs. Many, and I'm thinking about many of the so-called high-tech, specialty, or critical minerals and metals that are used in small quantities and yet provide essential properties or functionality to modern engineered materials, things like lithium and cobalt in batteries, neodymium and dysprosium in magnets, gallium and indium in electronics in flat-panel displays, and a variety of other applications.

Both research challenges apply here, chemical separations and resource efficiency, when it comes to improving and enhancing recycling of and recovery of minor metals from waste products.

With respect to recycling end-of-life products, as opposed to manufacturing wastes, demand for metals, almost certainly, will grow because of population growth, economic development, the lifting of many of the poorest people around the world out of poverty, and the improvement of their material well-being. Recycling by itself will not be able to meet this new demand because the quantities available for recycling today reflect the level of demand in the past. This is not to minimize the importance of enhanced recycling but rather to be cautious about the ultimate role of recycling in meeting our supply chain challenges.

So, as I began, government plays an essential role in fostering domestic supply chains of raw materials through research and education.

Thank you very much.

[The prepared statement of Dr. Eggert follows:]

Hearing to examine the United States' increasing dependence on foreign sources
of minerals and opportunities to rebuild and improve the supply chain
in the United states

Statement of

Roderick G. Eggert
Viola Vestal Coulter Foundation Chair in Mineral Economics
Division of Economics and Business
Colorado School of Mines

Before the

U.S. Senate Committee on Energy and Natural Resources

March 28, 2017

Roderick G. Eggert
Colorado School of Mines
March 28, 2017

Introduction

Chairman Murkowski, Ranking Member Cantwell, and Members of the Committee, thank you for the opportunity to speak today. I am Rod Eggert, Viola Vestal Coulter Foundation Chair in Mineral Economics at Colorado School of Mines. As part of my university responsibilities, I am deputy director of the Critical Materials Institute, an energy innovation hub (multi-institutional research consortium) funded by the U.S. Department of Energy and led by the Ames Laboratory. My area of expertise is the economics of mineral resources and materials.

I organize my remarks into three sections. First, I describe the context for current concerns about dependence on foreign sources of minerals and improving U.S. supply chains. Second, I present my views on appropriate roles for government in light of these concerns. Third, in the bulk of my testimony, I comment on the roles of research and education in fostering innovation and domestic supply chains for mineral resources and materials.

Context

First, it is not import dependence itself but rather *risky* import sources that are threats to U.S. users of mineral resources and the technologies that these resources underpin. In fact, import reliance is good if foreign sources are available at lower costs or are of higher quality than alternative domestic sources. In many cases, imports are simply intra-company transfers within a vertically integrated company; import reliance reflects an efficiently organized supply chain in which each step takes place in the location best suited to undertake this step. Approximately 62% of all U.S. imports, not just mineral resources, are intermediate products that U.S. entities use as inputs into the production of goods produced within the United States.[i]

Import dependence is a problem, however, when it puts supply chains and U.S. companies and material users at risk. Such is the case when imports come from one or a small number of production facilities, companies or countries – especially countries in which political decisions, restrictions on international trade, civil disruptions, or other developments may restrict access to materials for U.S. users.

Second, import dependence is one aspect of the broader and more-fundamental issue of supply-chain risk and raw-material availability. Short-term supply-chain risks may be due to: a limited number of mines, production facilities or companies (whether domestic or foreign); rapid, unanticipated demand growth for a material with small, existing markets; or reliance on by-product production of a material. Over the longer term, raw-material availability reflects: fundamental geochemical abundance or scarcity of chemical elements; investments in basic science, mineral exploration, mine development and process engineering to enable extraction and recovery of elements from rocks and minerals, manufacturing wastes and end-of-life products; environmental and social issues associated with mining lower-grade raw materials in more-remote locations; and availability of scientists, engineers and other professionals in the disciplines necessary for assuring material supply chains.

Third, the overall need for mineral resources will grow over time. Thus, existing sources and recycling will be insufficient to satisfy future demands (Ali, et al., 2017).

Government's Role[ii]
We appropriately rely primarily on private initiative to develop the mineral resources, materials and technologies that underpin today's society – technologies that encompass energy, health care, electronics and communications, transportation, environmental protection and national defense, among others.

But government plays essential roles in both establishing the institutional framework in which private activities occur and acting when markets do not work well. With respect to mineral resources and raw-material supply chains, government plays essential roles in:

- Facilitating undistorted international trade,
- Establishing a framework for efficient development of domestic natural resources that appropriately protects the natural environment and considers not only national needs but also the interests of the communities in which resource development occurs,
- Collecting and disseminating information, as well as carrying out strategic analysis, on which both private and public decisions can be made, and
- Fostering innovation and domestic supply chains through research and education.

The first role is outside the scope of this hearing. The second and third roles are the subject of other testimony at this hearing. Thus, I focus the rest of my testimony on the fourth role, fostering innovation and domestic supply chains through research and education.

Fostering Innovation and Domestic Supply Chains Through Research and Education
Although not a panacea, innovation is key to improving human living standards, environmental quality and even social well-being. Research and education are the means through which innovation occurs.

Private companies and individuals certainly have incentives to, and do, invest in research and education because of the benefits they bring to companies and individuals. But from society's perspective, private companies and individuals by themselves underinvest in research and education because the benefits are uncertain, often far in the future and often difficult for companies and individuals to fully capture.

Research. Over the longer term, there are three fundamental ways to manage supply-chain risks and assure resource availability: (1) enhance and diversify production, (2) waste less, and (3) use less. Research creates knowledge and technological options in all three areas.

Innovation to enhance and diversify production is the domain of research in basic geoscience, mineral processing and extractive metallurgy.

Innovation to waste less is the domain of research in improving manufacturing efficiency and increasing recycling of both manufacturing wastes and end-of-life products.

Innovation to use less, especially of those materials that are subject to the greatest short-term supply chain risks and long-term constraints on availability, is the domain of materials science and engineering.

Among the grand research challenges central both to enhancing and diversifying production and to reducing wastes are:

- Chemical separations, as highlighted by a 2016 paper in *Nature*, which identifies improving the separation of rare-earth elements as potentially revolutionary in terms of unlocking new and greater quantities of resources using less energy and with less environmental damage (Sholl and Lively, 2016), and
- Resource efficiency, enhancing the degree to which we recover multiple elements and materials that exist in a mineral deposit, manufacturing waste or end-of-life product (Söderholm and Tilton, 2012; Eggert, 2016). In practice, mining and recycling operations appropriately are driven by commercial considerations. These operations optimize the recovery of the most-valuable element or elements, which typically comes by not recovering any or all of the less-valuable elements that might be recovered. Innovation has the potential to improve the technical efficiency of recovery and to lower processing costs.

There are special roles for government to play in two specific aspects of research:

- Facilitating early-stage research and development (R&D) that is especially prone to underinvestment by the private sector acting alone, for reasons described above, and
- Facilitating the commercialization of promising ideas and new knowledge created in early-stage R&D through mechanisms such as public-private partnerships. In a perfect world, any promising new idea developed at a national laboratory or university would be picked up by the private sector. In practice, however, promising ideas often languish because of insufficient communication between basic researchers and commercial developers of new technologies.

Education. Education and research go hand in hand. Educational programs, especially those at the graduate level, educate and train the next generation of scientists and engineers, who in the future will respond to concerns about newly emerging critical minerals. Education and research in the geosciences, mining, mineral processing and extractive metallurgy, environmental science and engineering, manufacturing, and recycling can mitigate supply risks and increase material availability. Improvements in materials design—fostered by education and research in materials science and engineering—can ease the pressures imposed by those elements and materials subject to supply risks or limited availability.

Part of the educational challenge today is broad and relates to study of science, technology, engineering and mathematics, as highlighted in a number of National Research Council studies (for example, U.S. National Research Council 2007 and 2012).

Part of the educational challenge is narrower and relates to discipline-specific issues and the dearth of professionals in economic geology, mining, mineral processing and extractive metallurgy. A 2013 study of the National Research Council highlights these issues (U.S. National Research Council, 2013). Without well-educated professionals in the necessary disciplines, it will be difficult to rebuild and improve raw-material supply chains in the United States.

Critical Materials Institute. One example of an existing federal activity in the area of innovation and raw-material supply chains is the Critical Materials Institute (CMI), an energy innovation hub funded by the Department of Energy (http://cmi.ameslab.gov). The special focus of this research initiative is developing technological options for assuring supply chains of materials that provide essential properties to emerging clean-energy technologies, including high-efficiency motors, batteries, advanced lighting and solar materials.

CMI conducts early-stage research in all three areas identified above: to diversify and expand the availability of materials throughout their supply chains, to reduce wastes by increasing efficiency of manufacturing and recycling, and to reduce demand by identifying substitutes for materials subject to supply-chain risks. CMI also facilitates the commercialization of the new knowledge it creates through the active participation of its industry members.

Among CMI's current priorities is demonstrating the production of NdFeB magnets, essential in high-efficiency motors and at present produced almost entirely in China, using materials and technologies located entirely in the United States.

Recycling. Domestic supply chains already are well established for recycling manufacturing wastes and end-of-life products containing the major metals used in construction, transportation equipment, consumer durables and capital equipment – especially steel, aluminum, copper and lead. On the other hand, relatively little recycling takes place that recovers minor and specialty metals from end-of-life products. These minor and specialty metals typically are used in small quantities but provide essential properties and functionality to modern engineered materials – for example, neodymium in permanent magnets used in high-efficiency motors, lithium and cobalt in batteries, yttrium and europium in fluorescent lighting, and germanium and indium in flat-panel displays.

Considerable research is ongoing at present, including in CMI, to develop processes that will improve the technical and commercial attractiveness of recovering these minor and specialty metals. The technical challenges of separating and recovering multiple minor elements from complex materials are considerable – the grand challenge of resource efficiency that I noted above. But we are optimistic that with time and effort these challenges can be overcome. There are two other considerations, however, that lead me to be cautious about how large recycling's role will be in supplying these minor and specialty metals.

First, products containing these elements often are widely dispersed when they no longer are used – think of old cell phones, computers, computer monitors and television sets, which often wind up in desk drawers, attics and basements. The degree to which used aluminum cans were recycled fell with the spread of single-stream recycling and the demise of reverse vending machines. Without better social systems for collecting the products that are potential sources of minor and specialty metals, recycling will be limited.

Second, and more importantly, demand is likely to grow significantly for products containing these minor and specialty elements, and these products have lifetimes that range from years to a decade or more. The faster the rate of demand growth and the longer the product lifetime, the lower the percentage of demand that can be satisfied through recycling of end-of-life products. Consider a simplistic example with the elements silver, indium, and tellurium that are minor (but essential) constituents in several types of solar materials. A typical solar panel is expected to last twenty years or more. Imagine that (a) 10 units of a minor element are contained in solar panels installed this year, (b) over the life of these solar panels, the demand for the solar panels triples and (c) as a result the demand for these elements increases to 30 units per year. Future recovery of these minor elements when today's solar panels are recycled at most could satisfy one-third of the future demand for these elements, assuming no loss of material during recycling.

I am not suggesting that recycling is not an important focus of R&D efforts; recycling R&D is essential. Rather I am urging us not to think of recycling as a major substitute for resource development and mining. Both recycling and new mines will be required to meet future demands. Innovation through research and education is key to rebuilding and improving domestic supply chains of minerals and materials.

Closing
Thank you for the opportunity to testify today. I am happy to address any questions the Committee Members have.

References

Ali, S.H., D. Giurco, N. Arndt, and others. "Mineral supply for sustainable development requires resource governance," *Nature*, volume 543, 16 March 2017, pp. 367-372. Available at www.nature.com.

American Physical Society and Materials Research Society. *Energy Critical Elements: Securing Materials for Emerging Technologies* (Washington, D.C., American Physical Society, 2011). Available at www.aps.org.

Eggert, Roderick G. "Critical Minerals and Emerging Technologies," *Issues in Science and Technology*, volume XXVI, number 4, 2010, pp. 49-58. Available at www.issues.org.

Eggert, Roderick G. "Minerals Go Critical," *Nature Chemistry*, volume 3, September 2011, pp. 688-691. Available at www.nature.com/nchem/journal/v3/n9/index.html.

Eggert, Roderick G. "Economics perspectives on sustainability, mineral development, and metal life cycles," pp. 467-484 in Reed M. Izatt, editor *Metal Sustainability: Global Challenges, Consequences, and Prospects* (Hoboken, New Jersey, Wiley, 2016).

Sholl, David S. and Ryan P. Lively. "Seven chemical separations to change the world," *Nature*, volume 532, Issue 7600, 28 April 2016, pp. 435-437. Available at www.nature.com.

Söderholm, P. and J.E. Tilton. "Material efficiency: an economic perspective," *Resources, Conservation and Recycling*, volume 61, April 2012, pp. 75-82. Available at http://www.sciencedirect.com/science?_ob=ArticleListURL&_method=list&_ArticleListID=1170016130&_sort=r&_st=13&view=c&md5=649aaf21c8a33c30dadf7241db3e21a0&searchtype=a.

U.S. National Research Council. *Rising Above the Gathering Storm: Energizing and Employing America for a Brighter Economic Future* (Washington, D.C., National Academies Press, 2007). Available at www.nap.edu.

U.S. National Research Council. *Minerals, Critical Minerals, and the U.S. Economy* (Washington, D.C., National Academies Press, 2008). Available at www.nap.edu.

U.S. National Research Council. *Rising Above the Gathering Storm: Developing Regional Innovation Environments* (Washington, D.C., National Academies Press, 2012). Available at www.nap.edu.

U.S. National Research Council. *Emerging Workforce Trends in the U.S. Energy and Mining Industries: A Call to Action* (Washington, D.C., National Academies Press, 2013). Available at www.nap.edu.

[i] Calculated with data from https://jgea.org/resources/jgea/ojs/index.php/jgea/article/view/23

[ii] See Eggert (2010) and Eggert (2011), as well as two expert-panel reports in which I participated (American Physical Society and Materials Research Society, 2011; U.S. National Research Council, 2008). My testimony today on government roles includes views I expressed in previous testimony before (a) the Subcommittee on Energy, Senate Committee on Energy and Natural Resources, September 30, 2010, (b) the Committee on Industry, Research, and Energy of the European Parliament, January 26, 2011, (c) the Subcommittee on Energy and Mineral Resources, House Committee on Natural Resources, May 24, 2011, and (d) the Senate Committee on Energy and Natural Resources, January 28, 2014.

CHAIRMAN. Thank you, Dr. Eggert.

Thank you all. We have heard very interesting comments this morning.

I am going to begin with just a general question to whomever may want to jump in or multiples of you. This is the fourth hearing we have had in this Committee on the issue of mineral security.

A couple of you have testified before the Committee before. We have heard from USGS before. Three of you have flown in from other countries. You are clearly paying attention to the situation here in the United States. Other countries are paying attention to this issue. I think, most notably, China.

But here we are, and the information that you have given me this morning is that instead of lessening our dependence, we are actually increasing our dependence. We have increased it from just last year. We are not making headway on this issue.

It is a little bit frustrating, maybe because I feel like I am a voice in the wilderness sometimes here on these issues, but I have been trying to raise the issue, raise the profile, speak to what it means when we are more vulnerable or relying more on risky imports, to use the terminology that was given here today. What are we doing wrong here?

This smartphone that you all have in your pockets, that you are using to take pictures, it does not happen without these critical minerals. Those of you that flew here would never have been able to arrive had we not had these.

So much of this is education, education, education. I think it, kind of, fits with my view of how many people, how many in this country, view energy in general. There is this immaculate conception theory of energy. It just happens.

I am starting to think that same view holds true when it comes to how we are able to operate as a society. We do not make the connection to where our minerals fit in. What can we be doing more to make this connection?

Dr. Hitzman, you mentioned the fact that to this point in time only one-third of the United States has been mapped. We clearly have some room to grow there.

But from the perspective of educating, whether it is our manufacturers, who are part of that supply chain so I think they get it. But do we, as a society, get it?

It is one thing when you mentioned that we are impacted by the ability to get a red car or a black car because the Chinese acted and cutoff those rare earths there. I don't think people get too alarmed about the fact that they might not be able to get the color of their choice. But when they view that this is a security threat, that changes the discussion, one would think.

I am kind of throwing this out there for general discussion. What are we? Where are we failing to connect with Americans, not only John Q. Public out there, but folks in the White House as well? How do we raise this up beyond just this Committee?

I welcome anyone to comment. Dr. Hinde?

Dr. HINDE. I'd love to say I had an answer, but as I——

CHAIRMAN. I was hoping for it.

Dr. HINDE. I've been writing about this issue for 30 years. In fact, I launched an environmental magazine about 15 years ago to address these very issues.

I mean, it stems, of course, to state the blindingly obvious, from a mistrust of the industry. I can't speak for here, but certainly in Europe, we were pretty bad miners in the last century and we were awful the century before that. Even the Romans didn't mine terribly responsibly.

So throughout Europe we've got historical baggage. We've got some pretty shocking lignite, remains of lignite, mines. And so, most Europeans certainly grow up with a dislike, inherent dislike, of the mining industry.

I'm a mining engineer, but neither of my sons went into mining, both went into accountancy. We have a serious issue facing the industry because at the school level, it's not understood.

I think it's probably more serious than even you've painted it in that it's not just the link between metals and what we use. That should be doable. I mean, the popular, certainly in North America and Europe, should understand these things. They might choose not to notice.

The more serious thing is just not getting mining. They're quite happy for it not to be in their backyard. They want someone else to do the hard yards and make the metal. As we've elucidated here, that isn't a very clever strategy for the future in terms of security of supply.

But if you can have your products and someone else does the digging, that looks preferable to most people at the moment. So, that's not an answer to how to solve it, but it's clearly got to start at the school level, that responsible mining is a way forward. It just has to be done environmentally in a friendly manner which we are now doing.

CHAIRMAN. Yes, I appreciate that.

Mr. Barrios?

Mr. BARRIOS. I would say also in terms of storytelling and comparing to other countries. If we compare the permitting process in the U.S. versus Canada, we can clearly see that in terms of scope and depth, the permitting process in Canada is very similar to the process in the U.S., the consultation process, the amount of rigor and discipline that goes into the process.

I think talking about how the people are doing it and trying to address the issues that are becoming obstacles, to be able to be as effective as other countries in allowing mining projects to progress at an acceptable speed. And I think the timeline, what I mentioned before, is critical.

If you look at the process in Canada, clearly the timeline is very different. A number of colleagues mentioned it. I mean, it's truly about being rigorous and disciplined with the amount of time that one is assigning for these permitting processes to take place. And it is important for companies like ourselves and other mining companies. If there is one thing which we're looking for is certainty. Clearly, that lack of certainty in the timeline does impact our ability to be able to put forward projects in the U.S. and make them as competitive as projects in other parts of the world.

CHAIRMAN. Well, I appreciate that. I have, kind of, thrown it out to all of you. My time has expired, but if we want to come back to visit at the end of the hearing, if any of you have additional comments you want to add to that, I would welcome that.

Senator Cortez Masto?

Senator CORTEZ MASTO. Thank you, Madam Chair.

Let me follow up on this discussion on permitting because, as a new member to the Committee and somebody who is from a state that grew up with mining in our state, this is something I have constantly heard is the permitting process impeding, really, the movement forward when it comes to mining. I constantly hear it, but I don't hear specifics. Now I am sure our Chairwoman or Ranking Member and many others are focused on this.

Can you give me an idea, when we are talking about a permitting process that has taken seven to ten years, what is it, specifically, that we can do at the federal level to streamline it or are there duplicative processes that I have heard from Dr. Hinde as well? What is it, specifically, that we can focus on to cut that time down to address what I have heard today from all of you?

Dr. HINDE. I don't pretend to be an expert on USA permitting but we've, obviously, done quite a lot of surveys asking other people's opinions, but there were two primary differences.

In Canada and Australia, for example, and certainly at the federal level there, they're also not coordinated. It's at the state level the difference comes in. Broadly speaking at the state level in Australia and Canada, one agency takes the lead. They set the goals, they set the timeframes and other agencies link to them. And in that way, they try and avoid overlapping requirements. The total requirements are no more rigorous. They're very similar, but what they do is they set the benchmark for other people to do and generally speaking, they hit the time tables.

The second thing that is different is that in both those countries it is the mining company that does the environmental impact statement (EIS). They obviously use third parties. They use consultancies that, I think, can be relied upon, but the company pays for it and organizes it and does the timeframe. Of course, it's in their interest to drive it. If it's left as it is here with an agency to set the environmental impact statement, there isn't quite the same urgency. Clearly, the agency needs to monitor and make sure that EIS has been done, done adequately. More often than not, it's done by an international consultancy company, whoever it is that's tasked them with the requirement.

Senator CORTEZ MASTO. Thank you.

Mr. Barrios, I am curious, any specific thoughts on how we can streamline it or concerns?

Mr. BARRIOS. I think, similarly, I mean, when we look at it from a Canadian perspective, that my colleague mentioned as well, but I would really highlight if one looks at Canada the standards are very similar.

It's about the timelines. It's about making sure the roles and responsibilities of each agency and the timeline base targets are agreed and published at the start of the application process so we all know what the timelines are and those are adhered to. And that, really, is one of the key elements that is making a difference in the permitting process where we're finding that in two to three years you can obtain them in Canada. It's been lengthening here in the U.S. from five to seven, now to seven to ten. And this really is hurting investment.

Senator CORTEZ MASTO. Thank you.

Dr. HINDE. Can I, sorry?

Senator CORTEZ MASTO. Please.

Dr. HINDE. Can I just add to that?

The one thing I forgot to mention that we did find in our survey of a year and a half ago, was that here, unlike in Canada and Australia, sometimes the same requirement can be repeated over rather than sit down in the beginning and hear from the various interested parties what is it you need to test or check and put it together in one document and do it in one go.

The mining companies here, to a certain extent, are asked to do one particular environmental impact assessment and then perhaps six months later someone else chips in and it's oh, I would like to do something slightly different and they do it again.

Far better, clearly, to get it all done in one go, even if it's more rigorous at that point and takes longer. It's parallel permitting as opposed to in series.

Senator CORTEZ MASTO. Thank you.

I know my time is running short, but Dr. Eggert, I am curious, your thoughts on this?

As you well know, besides a school like yours, Nevada also has a College of Mines. I know that approximately 70 percent of mining engineers will retire within the next decade. And because fewer and fewer students are enrolling in mining engineering programs, we will not be able to replace them at an adequate pace.

What recommendations do you have to increase enrollment of students in these programs so that we do have a robust workforce?

Dr. EGGERT. I think one of the key actions that would help improve enrollments in mining engineering, mineral processing and extractive metallurgy is actual research funding in this area that will allow faculty members in these departments to hire graduate students.

Senator CORTEZ MASTO. Thank you.

Dr. EGGERT. I mean the single thing that I would suggest.

Senator CORTEZ MASTO. Thank you, I appreciate it.

I know my time is up. Thank you, Madam Chair.

CHAIRMAN. Thank you.

Senator Daines?

Senator DAINES. Thank you, Madam Chair, for having this hearing today. This is very important for my home state of Montana.

I do want to thank the Committee and the witnesses today for highlighting the importance of critical materials for the United States and the very high hurdles we have to jump over to extract them.

What too many people forget, and the Chair mentioned this in her opening comments, is that if the U.S. wants to continue to be a leader in high tech, in communications, renewable energy, we have to be a leader in critical mineral development. Everything from our cell phones, telephone lines and wind turbines require these critical minerals.

In my home state of Montana, mining is a backbone, so much so that it is written into our state motto, "Oro Y Plata," gold and silver. If you look at the Montana State flag, it says Montana on it and then there's Oro Y Plata. Those are the only words on it. They are in Spanish. In fact, you will see a shovel and a pick axe there next to a plow, going back to the very roots and the foundation of our state of agriculture as well as mining and natural resources.

The Still Water Mine in Montana is the only, let me say that again, is the only producer in the United States of platinum and palladium, the only one.

We are a major copper producing state as well.

At the same time, Montana has received awards for our first-class reclamation work. Most Montanans are passionate about fly fishing and hunting and the outdoors and preserving the incredible, pristine environment that we have in Montana. And count me in on that.

At the same time, we must continue to responsibly develop our resources so that moms and dads can still stay there, raise their children there, and still go to Walmart to buy an elk tag, so we do not turn into a land only for the rich and famous because we do not have jobs there that working families need to have a living wage. These jobs, the mining industries, provide that. We are only producing in Montana about one percent of our potential, so there is a lot there.

We can begin to expand our critical mineral production by streamlining and speeding up the permitting process that was talked about here in your testimonies. The U.S., as was mentioned, has one of the longest permitting processes in the world. I will give you a couple of examples.

In Montana, we have the Rock Creek and the Montanore projects. They have been in the permitting process, now I heard seven to ten years, we would be envious of that kind of result. The Montanore and Rock Creek projects have been more than 30 years in the permitting process, and they are still not up and running. Do the quick math. Go back 30 years. Ronald Reagan was President. It seems like irony that we now have statues of presidents in Statuary Hall that were serving when the permitting process began some 30 years ago.

Here is the impact for families in Montana. The Forest Service estimates the Montanore Project would provide full-time employment for 450 people. The Rock Creek Mine will provide more than 300 full-time jobs. That is $667 million in direct payroll over the life of the project, and $175 million in tax revenue.

I can tell you, I spent a lot of time talking to my county commissioners back home, and they are struggling to find ways here to make ends meet from a tax base viewpoint. The indirect economic benefits are even greater than that.

By the way, these projects are in Lincoln County. It is a county in my state that has one of the highest unemployment rates. They can benefit greatly from this. I spoke to a couple a few years ago from Eureka, Montana, in Lincoln County and they said, "Steve, basically what we have in Lincoln County now is poverty with a view." We need to change that.

Mr. Barrios, in your testimony you speak about the length and the duplicity of the permitting process. Could you expand your suggestions to simplify the process? I know you had somewhat a similar question before. Maybe specifically, what can this Committee do? What would you recommend to us in terms of action we can take here to try to streamline the process?

Mr. BARRIOS. I think when you look globally at what are the overarching themes that a company like ours looks at when it is thinking about investment, it really is around regulatory certainty and it's in three areas. The reliable timeline of the permitting process, the second thing is creating certainty in access to minerals, and the third thing is finally having something that is reasonable around financial assurance, closure.

If we look at the timeline, I think that's where we emphasize that's one of the critical elements that we need to ensure that, similar to what we have in Canada, there are set lengths that are adhered to.

If we look at our Resolution Copper Project in Arizona, we started the permitting process in 2013. We've spent so far $1.3 billion, and we're far from completing the process there. This is a mine that will supply, could supply, 25 percent of the U.S. copper needs, and create 3,700 jobs. It's quite staggering that now in another country like Canada, we would be having those permits in our hands and processing—progressing with the project. We are still, through the process, trying to obtain those permits.

Senator DAINES. Thank you. I am out of time, but it sounds like our neighbors to the north may have some examples of, perhaps, processes and some parameters that may be helpful for us here.

Thank you.

CHAIRMAN. Thank you, Senator Daines.

Senator Stabenow.

Senator STABENOW. Thank you, Madam Chair, and thank you to each of you.

Coming from a state like Michigan, where particularly in Northern Michigan in what we call the upper peninsula, which has been mineral rich for a century, when watching things change there based on mining and having jobs and then not having jobs, I certainly understand the economic impact of what is being talked about.

Looking at your testimony I know that you are talking significantly about permitting issues and regulatory barriers impacting the industry, but I would like to talk for a moment about the importance of transportation infrastructure in all of this.

In Northern Michigan in the upper peninsula in Sault Ste. Marie we have a lock and dam that is vital to transporting mining goods, including iron ore, throughout the Great Lakes region and the country. According to the report by the Department of Homeland Security, a shutdown of the Sault locks would likely result in all North American production of mining equipment and automobiles and farming equipment to stop within weeks.

We have a very old infrastructure there, only one of the locks is big enough to handle most of the cargo going through there. I think we are on borrowed time at the moment with that lock.

Eleven million people would become unemployed if that lock shut down, even for a few weeks, and the North American economy would enter a severe recession.

I wonder if each of you might speak to how important it is from a mining industry standpoint to have well-functioning locks and dams, roads and bridges and rail to operate efficiently and compete in the global marketplace? And what does our aging infrastructure mean for our ability to move minerals and materials where they need to go?

I guess I will start at the end, yes.

Dr. HITZMAN. Thank you, Senator.

In terms of the USGS, we're not so much looking at the infrastructure, we're looking at where to get the minerals. Michigan, most people don't know, was actually the major supplier of copper to the world for a number of years.

Senator STABENOW. Yes, that is right.

Dr. HITZMAN. Clearly in any area of the world or the country where we're going to do mining, one of the things that the companies look for is sufficient infrastructure to actually move materials and then the mine products out. So it's clearly a critical part of the equation.

Senator STABENOW. Mr. Barrios?

Mr. BARRIOS. I think, similarly, one has to look project by project. It's very difficult to give a general answer. Generally it really depends where the resource is and how far it is to get it to market. So it is a critical element, and it makes a big difference in the evaluation of a project. That's usually, the transportation costs, are a significant cost of exploration. So it is a very critical, important element. But it really depends, resource-by-resource.

Senator STABENOW. Dr. Hinde?

Dr. HINDE. Yeah, the important part, I think, of infrastructure is to recall that infrastructure is absolutely crucial for bulk commodities, such as in your state, Senator. Clearly coal, copper and those big, bulk commodities, railway lines and infrastructure and ports are absolutely required.

But of course, half the mining industry, in terms of expenditure, is gold and that you can fly out by helicopter. It's less required for infrastructure, so it rather depends, as my colleague said, on a project-by-project basis.

The other thing to bear in mind is the USA constantly rates right at the top in terms of infrastructure on a world perspective. We all know, in this room, that your infrastructure is aging and needs work. But on a world perspective, it is highly regarded. And so, companies come here because of your infrastructure, notwithstanding your problems.

And so, you know, there are other things that are damaging the industry here like permitting rather than infrastructure.

Senator STABENOW. Well, it is interesting though being in China and being in Brazil and other places where they are putting large amounts of money into infrastructure. At some point, they are going to be ahead of us because we have not been doing that.

Dr. HINDE. Indeed, yeah.

Senator STABENOW. Yes, so—yes, Mr. MacGillivray?

Mr. MACGILLIVRAY. The only thing I could possibly say is that in Alaska we are actually looking for the roads in the first place. [Laughter.]

Sort of, roads to resources is our common theme that Madam Chair has been a proponent of. So, from our perspective, you have a good problem that you are able to readily access your resource base.

Senator STABENOW. Vice Admiral?

Admiral COSGRIFF. If the Chair will indulge me, thank you for asking a question about ships.

Senator STABENOW. Yes.

Admiral COSGRIFF. But if you're going to move something like an ore or heavy, dense commodity, then you'll want to move it on water. And if you can't get it out on water that flows, you'll want it in a pipeline and if it doesn't flow, you'll want it on a train and so on down the path.

At the far end of this process we've received these materials largely over road, rail and road, and then when we finish our jobs as manufacturers, they go out the other side on, principally, road and rail.

This full scope look at our infrastructure is, in our opinion, long overdue. It, in and of itself, is an investment in real estate or in infrastructure, transportation infrastructure, along with other types, like electrical, which will pay dividends for this country over the longer run.

Senator STABENOW. Thank you.

Dr. Eggert?

Dr. EGGERT. Yes, I agree with what others have said. Infrastructure, in general, is important for mining and other forms of economic activity. With respect to mining, it's especially important, as Dr. Hinde said, for the bulk commodities.

Senator STABENOW. Thank you, Madam Chair.

CHAIRMAN. That is a great question, and it is so key to everything.

As Mr. MacGillivray says, we have got the resources there but we do not have any way to get to them or get them out. So infrastructure is key and certainly something that this Committee has been focused on of late.

Senator King.

Senator KING. Thank you, Madam Chair.

Some specific questions.

First, Mr. MacGillivray, what is the nature of rare earth mining? In other words, is it tunnels, pits, mountain top removal? What are we talking about here in terms of how it is actually, physically, done?

Mr. MACGILLIVRAY. So the nature of the deposits do vary. There are proposed projects in the United States that are open pit but our project in Southeast Alaska is a vein-hosted deposit; therefore, it would be accessed by underground methods.

Senator KING. So it varies? It varies according to the deposit and where it is?

Mr. MACGILLIVRAY. Based on the geological occurrence.

Senator KING. Are there any special environmental problems associated with these particular minerals as compared with coal or oil or gas?

Mr. MACGILLIVRAY. No, I don't think there's anything unique with rare earth deposits. Maybe there is, you know, some slight enrichment in uranium and thorium that has to be considered and dealt with appropriately, but by and large they're similar to other commodities.

Senator KING. Mr. Hitzman, do we have rare earths, significant deposits of rare earths, in the United States if we could do the development necessary?

Dr. HITZMAN. I think you're hearing from one of our panel members who has one. So that's one, and there are others that companies are working on in various parts of the United States, Wyoming and of course, the large deposit in Southern California that has gone in and out of production. So, the answer is yes, we do have deposits.

Senator KING. Is there more, is there potentially more, if we had better mapping and geology?

Dr. HITZMAN. Absolutely.

Senator KING. I know the Chinese, for example, are buying up mines and resources around the world, not necessarily in China, but they are buying properties in Africa and South America. Is that, are our mining companies doing something similar? Are we looking all over the world for these materials?

Somebody?

Dr. HITZMAN. I can answer that from USGS.

Yes, American mining companies are exploring around the planet. Just like Rio Tinto which is a major, multi-national company working all over the world. Freeport and other companies in the United States, Newmont, are also doing the same.

Senator KING. Okay. We have talked about the fact that we are dependent. I commend the Committee's attention to the chart the Committee staff included that is really pretty shocking that shows—we are 100 percent dependent on 21 minerals from other countries, which is a dangerous place to be, particularly when they have strategic value.

What is the bottleneck? I know you have talked about permitting. It sounds like we have a loss of engineers; we have financing issues, in part relating to permitting; we have permitting; and, we have fundamental geological research. Is that a good list of what the obstacles are? Does somebody want to echo that?

Yes, sir?

Dr. EGGERT. Yes, that's a reasonable list. It's, I would say, not a single factor, but a combination of several factors.

With respect to rare earth resources, in particular, there are special technical challenges associated with separating the rare earth elements from one another.

Senator KING. Does that have to happen at the mine or can it be shipped somewhere else with the separating happening somewhere else?

Dr. EGGERT. Typically what happens is that the mineral resource is concentrated at the mine site and then often, initial separation.

There are 15 or so rare earth elements and the initial separations involve separating them into, basically, two or three piles.

Senator KING. Okay.

Dr. EGGERT. And then there are subsequent separations that can take place at the mine site or elsewhere.

Senator KING. I want to talk a bit about permitting, and I know I am running out of time.

Quick question. If federal lands are used for one of these mines, are there royalties paid to the taxpayers for the extraction? Mr. Hitzman?

Dr. HITZMAN. Yes.

Senator KING. Okay, so there are royalties that come back for whatever the value is of the mined minerals.

Obviously, Madam Chair, we have got to talk a lot about permitting. I would like to know, specifically, where the bottlenecks are in the permitting. And is it a lack of deadlines, is it multiple studies, is it multiple agencies?

In Maine we had these issues and we, in part, solved them by having a lead agency where it was a one stop permitting. The lead agency would coordinate the studies that were necessary.

I am getting a lot of nods. Is that a——

Dr. HINDE. Yeah, that's exactly the issue and that's what they essentially do in Canada and Australia, somebody takes the lead and organizes all the other interested parties.

Senator KING. I take it that does not happen here? You have got to get 27 separate permits.

Dr. HINDE. It would be, appear to be, the exception rather than the rule.

Senator KING. So that is something, Madam Chair, obviously, we want to look at.

My final question is for you, Madam Chair. Are you going to reintroduce S. 883, or have you?

CHAIRMAN. From this wonderful hearing I plan on reintroducing it if we need to add anything, but the purpose of the hearing was designed to help us, kind of, supplement that, if necessary. So yes, I am intending to reintroduce S. 883 and would welcome the support from other colleagues.

Senator KING. Well I would like to work with you on this because, based upon my service on the Armed Services and Intelligence Committees, this is a national security issue and I think we need to find ways to have a predictable and timely permitting process that still adequately protects the environment. So I would like to work with you on that.

CHAIRMAN. I appreciate that. Know that I absolutely concur in terms of the security perspective. It is something that we need to be working on, so I appreciate that.

Senator Barrasso.

Senator BARRASSO. Thank you, Madam Chairman.

Dr. Hitzman, soda ash producers in Wyoming, like so many others in the minerals industry, face increasing transportation costs, as well as intense competition from foreign markets. The cost to ship soda ash from rural Wyoming to ports and domestic consumers is substantial. So foreign suppliers are able to subsidize their production and do not face many of the regulatory overheads that the suppliers in the United States face.

In your view, what can Congress do to ensure a strong, domestic market so that American producers are able to remain competitive?

Dr. HITZMAN. Well, it's, sort of, many different things, not one individual thing. One is ensuring that the transportation infrastructure exists to help get things to market. Ensure that various parts of the tax code work to the benefit. That's something that's coming up. And actually, ensure that producers have, as other people have said, certainly with how the laws are applied to the minerals industry.

Senator BARRASSO. Nearly all of you on the panel today have suggested in one way or another that the United States should reduce our reliance on imported minerals for either economic or national security reasons. Senator King just made that reference.

The other side of the coin is improving the ability to export raw materials and goods. You know, in Wyoming and in any other mineral producing state, our resource industries require access to foreign markets and you need to get through ports. It is becoming increasingly more difficult for these industries, I believe, to gain access to these ports.

Mr. MacGillivray, to your point, you discussed ongoing environmental issues with Chinese production of certain minerals that the United States also produces. So in your opinion, what steps can Congress take to improve trade pathways through coastal ports so that these cleaner, American-made, raw materials and goods have access to foreign markets?

Mr. MACGILLIVRAY. So in my answer I'd like to restrict my comments to rare earth materials, critical and strategic materials.

As Dr. Eggert correctly identified, the crux of the issue with production in the United States is the separation technology. It's the sole reason that China dominates the monopoly that they do with rare earth production right now because they have limited regard for the environment so they use a technology called, or a technique called, solvent extraction.

Ucore Rare Metals knew when we were permitting the Bokan-Dotson Ridge Project that solvent extraction would not be permittable in Southeast Alaska, an environmentally sensitive area, so we shopped the world for alternative technologies and came across a Nobel prize winning technology called molecular recognition technology. It's a technology that's not only limited to mining, it's also used in the healthcare industry. But the basis of it is ligand based, so there are no solvents. There are no extreme pollutants from this process. It's very innovative and adapted toward this issue.

So, I guess, a shorter answer here is that, some sort of support to help develop rare earth separation in the United States will enable us to have domestic supply and then be able to export, eventually, materials to other manufacturers worldwide.

Thank you.

Senator BARRASSO. Thank you.

In your written testimony, Dr. Eggert, you identified the need for the government to establish an efficient framework that both protects the environment and considers the needs of the community where the development occurs, and I agree.

Dr. Hinde, in your written testimony you mentioned that new mines can lose one third of their economic value as a result of delays in production, more than 30 percent of the value of a mine could be lost because of permitting delays.

In Wyoming we have one of the biggest reserves of rare earth minerals in the world, but companies face decades long permitting delays and tens of millions of dollars in up-front costs. So, I believe, now is the time that we should create some certainty in the job market and in national security.

Dr. Hinde, Dr. Eggert, can you just talk a little bit about how much certainty do you think addressing these unnecessary permitting delays would bring to the industry, and how do we eliminate these unnecessary and unreasonable permitting delays, especially those not caused by the applicants themselves?

Dr. Eggert?

Dr. EGGERT. I think Dr. Hinde made a couple of very useful suggestions, the appointment of a lead agency that establishes the framework and a timeline for the permitting process.

More generally, I think what companies are looking for is certainty in a process as opposed to certainty in actual outcomes. In other words, a process that gives them a fair hearing, you know, in what various parties, all parties, would consider to be a reasonable timeframe.

Senator BARRASSO. Okay.

Dr. Hinde?

Dr. HINDE. Yeah, essentially, exactly the same. I mean, almost across mining, it's certainty whether it's in tax or any sort of legislative and working environment. It is just certainty. Given that, we can plan accordingly.

Senator BARRASSO. Thank you.

Thank you, Madam Chairman.

CHAIRMAN. Thank you.

Senator Cantwell.

Senator CANTWELL. Thank you, Madam Chair, and thanks for holding this hearing.

I had a chance to chair a hearing a few years ago on critical minerals in general, so it is very important for us to continue our focus in this area.

Dr. Eggert, I wanted to ask you about recycling of critical mineral materials and what you think the recycling opportunities are for us, as it relates to supply?

One of the things we have been proud to do in the northwest, as we shift to composite manufacturing, is to look at recycling as a way to drive down the cost of composite materials for smaller businesses. I wondered what you thought about, as we look at shortage issues, looking at recycling of product too?

Dr. EGGERT. I think recycling has an important role to play, and its role can be enhanced.

As I indicated in my written and oral testimony, very little recycling takes place at present of the, so-called, miner or specialty metals that appear in small quantities and yet, provide essential functions to modern materials.

A key challenge, part of the challenge, is technological. Elements like indium in flat panel displays are there in very small quantities and therefore, the economic case is not going to be made on the basis simply of indium, but the ability to recover several materials.

The current technologies really focus on the major, most valuable, elements in a product and there's technical work to be done at, what I call, the resource efficiency, optimizing the recovery of multiple elements from a multi-element product, like a smartphone or a television set.

And it's really a similar set of issues to recovering multiple elements from a mineral deposit. Most mineral deposits contain multiple elements, only a couple of which are actually recovered for commercial reasons.

Senator CANTWELL. How do you think we could proceed in this area? I know some of our labs are doing work, and do you think the private sector just continues to——

Dr. EGGERT. Well, I think the private sector is doing work in this area.

A number of national labs are and in fact, I'm involved in an entity called the Critical Materials Institute which is a Department of Energy-funded research consortium that has as its members at universities, companies and national labs. It carries out early stage research related to, among other things, recycling of critical materials. Industry partners help us identify key challenges and important problems. And so, I think a continuation, perhaps an enhancement, of this type of public/private partnership that forces companies and national labs and university researchers to talk to one another, better than maybe they have in the past.

Senator CANTWELL. I personally like those models because you are then getting the maximum out of everybody at the table. I am very big, obviously, on collaborative efforts in general. So anyway, we'll look forward to discussing this with you further.

Thank you, Madam Chair.

CHAIRMAN. Thank you, Senator Cantwell.

Senator Lee.

Senator LEE. Thank you very much, Madam Chair. Thanks to all of you for being here today.

Mr. Barrios, you mentioned a shocking number, a significant number, that Rio Tinto has spent—some $1.3 billion on permitting studies, on permitting, on studies and on shaping the Resolution Copper Mine. Now, this is great. We love to see investment in these kinds of things. I am glad that you are able to put those resources into it and that you have access to resources that will benefit consumers in America and throughout the world.

My concern is that our current regulatory regime makes it very, very difficult for anybody to do anything. It basically prohibits mining investment from non-Fortune 500 companies. There are very few companies out there, very few people anywhere, who can afford this type of investment.

As if the current regulatory burden were not enough in this area, on January 11th of this year, the Obama Administration proposed a rule to create additional bonding requirements under section 108 of CERCLA for hard rock mining. If the proposed CERCLA rule were finalized, tell me, sir, what effect might that have on the mining industry and on your ability to extract critical minerals?

Mr. BARRIOS. Thank you, Senator Lee, for that question.

The CERCLA 108(b) for us is clearly a disincentive, but furthermore, I would say in terms of investment, because of the burden it implies, but the issue which is a concern is the duplication in terms of financial assurance at the state level and the federal level. This is really an area where we could see some simplification and avoid duplicating rules and regulations that are not adding additional value.

Senator LEE. I think everyone here agrees that mining companies and industrial producers need to be liable, need to be responsible for any disasters they create for superfund sites they create, that, of course, have to be cleaned up. So that is not in dispute. If a company goes bankrupt or if a company walks away from a contaminated site, the American people should not be faced with having to either foot the bill for the cleanup or, alternatively, face the catastrophic consequences associated with just leaving it there. What bonding requirements and regulations, state and federal, are currently in place to ensure that mining companies leave mining sites in a stable condition?

Mr. BARRIOS. The issue that we see, and I mentioned before, really is around the CERCLA 108(b) rule. It is an example of a regulation which is duplicative and unnecessary. We already see the current programs that are in place address the risk of mining and mining processing sites and prevents these sites from becoming a superfund liability. So for us, really, this renders the current rule being proposed unnecessary.

Furthermore, I think we can say with certainty that the practices that lead to contamination of groundwater, soil and wetlands in the past, simply are not allowed today under the many state and federal requirements that we must meet.

Senator LEE. So in your opinion those existing requirements obviate the need for these new regulations?

Mr. BARRIOS. Yes.

Senator LEE. Rio Tinto Kennecott has, of course, a long history in my state, in Utah. You have been operating in the Salt Lake Valley for over 100 years and plan to continue operating for a significant amount of time to come, and we are happy about that. But mining is not always easy. In 2013 the Kennecott mine suffered the mine slide which was very significant, and it was difficult.

Can you describe the recovery process and also other sustainability efforts you have in place?

Mr. BARRIOS. Yup.

Rio Tinto was aware of the slide potential in February 2013, and we began preparing for a safe and minimal impact event. We had nine layers of safety in place to monitor the material movement and safety was the number one priority at the time and it continues always to be at Rio Tinto.

We were very happy to report that nobody was injured during the event, and all the personnel were evacuated before the slide occurred.

We were also very proactive in engaging with the key external stakeholders prior to the slide, and the community was very appreciative of knowing the information beforehand. To this day, they still praise Rio Tinto for the transparency around this event.

The slide was a slide of 150 million tons which took place in the night of April 10th, 2013. The slide material would fill enough rail cars to stretch three-quarters the way around the world. It was quite a material slide.

The overburden we recovered very fast. The overburden was mined three days after the slide, and production started operating 17 days after the event. So very, very fast recovery. And we did spend about over a billion dollars to remediate the slide and materials. So quite a big commitment for the mine and to continue operating the mine for years to come.

Senator LEE. Thank you very much.

I see my time has expired. Thank you, Madam Chair.

CHAIRMAN. Thank you, Senator Lee.

Mr. Hitzman, I want to ask you where we are with the USGS budget and how much of your budget, the agency's budget, actually goes to the minerals work each year.

I am concerned that as we talk here today and try to shine a spotlight on things that we are not doing all that we need to be doing from an agency perspective, from the federal perspective, in making sure that we have the information, the data, the analysis, the mapping. Within USGS, how much time and how much of the budget actually goes to the minerals aspect of the work that the agency does?

Dr. HITZMAN. I actually don't know the exact percentage but it's not the largest of the mission areas in the Survey. It's one of the smaller mission areas.

The budget, over time, decreased for a number of years, but in the last couple of years has had a slight uptake and stabilized. Of course, now we're under a CR, so we're where we were last year.

CHAIRMAN. But as you have indicated to the Committee here one-third of the mapping that you believe that we need to have done as a nation, only one-third has been completed, so we obviously need to be resourcing this a little bit better. Is that a correct statement?

Dr. HITZMAN. It would be good to do that. Remember that not all the mapping is done through my part of the Survey.

CHAIRMAN. Right.

Dr. HITZMAN. As well.

CHAIRMAN. Right.

Dr. HITZMAN. So it's done through other pieces of the Survey.

CHAIRMAN. Right, okay.

You said in your testimony, in speaking about what was done with the Alaska mapping, recognizing what it is that we have allows us to then move out and do more. It allows those that are looking at it from an investment position to have a greater degree of certainty going forward. It seems to me that if we do not have solid mapping, it just further slows our process there.

Vice Admiral Cosgriff, as you represent those in the manufacturing industry, are you hearing concerns from your member organizations about the growing vulnerability that we have as a nation and that they have as U.S. manufacturers with the growing realization that we are relying more and more on imports?

Again, I think it was you, Dr. Eggert, you said it is not necessarily relying on imports so much as risky imports. But are you hearing concerns from your members about this issue and do you see growing pressure to see more action?

Admiral COSGRIFF. I don't think they would say it in that, sort of, global way you did about, sort of, a risk to the United States manufacturing, per se. It would tend to be more particular.

I can give you a good example though of how close that they watch where their supplies are coming from. You recall a few years ago a large-scale strike on the west coast which created the opportunity for a major disruption in supply chains coming, mostly, from the Far East. And so, the response to that was, as you'd expect, with things moving on ships you now have to find a different port for the ship to go to. You have to figure out where your inventory stocks are, for how many days of production you have left before that ship gets to wherever it's going to get to. What are your alternative sources of moving that input from, let's say, Long Beach, California, via rail to someplace on the east coast and then by truck to your plant?

So, that was a major event, and I think it served as a wakeup call for a lot of our companies to pay far more attention, even more attention, than they already are paying to the supply chain. I think, to some extent, the discussion about NAFTA is having a similar effect. That's a material effect on a supply chain, in this case, closer to home, a very mature supply chain, that again, has their attention and that we have to get right so that we don't disrupt those supplies.

The bottom line, though, it is a globally sourced supply chain still with the few exceptions we've talked about today that do have the attention of our company's rare earths, I put at the very top of that.

CHAIRMAN. Yes, I can remember when the Chinese effectively cut off all sources to Japan over a dispute with Japan. It certainly got the attention of those in Japan and, I think, those of us in this country as well because you realize then the real stranglehold, the chokehold, that China has when it comes to the rare earths.

Dr. Eggert, you have spoken a little bit in your testimony here today, as well as in your written testimony, about minerals research and we have discussed the mapping aspect of it. But what research are we seeing being conducted at our universities, at our national labs, the Critical Materials Institute, to make the mining, the processing, and the end use of critical minerals more economically viable? Are we seeing the level of research that you believe is necessary?

Dr. EGGERT. I'm not sure I can speak to the level of research. I guess my bias would be, as researchers, we would like a higher level of funding.

But I can describe what's happening using the Critical Materials Institute, this Department of Energy-funded research consortium that I mentioned earlier.

If you think about supply chain risks or long-term resource availability, there are really three solutions, and technology plays an important role in all three.

There's first of all, technology that enhances and diversifies production, technology that enhances or reduces waste, and technology that helps us use less. And so, it's process engineering in the first two cases and it is material science in engineering in the third case.

The Critical Materials Institute is carrying out research in all three areas. As I indicated in my written testimony, I think of the many research challenges, the two grand challenges, or at least two of the grand challenges, are chemical separations, which are important both for mineral resource development and production, and the recycling of manufacturing wastes and end of life products.

And so, these two types of research are really quite complementary in terms of both the chemical separations and the other one that I mentioned, resource efficiency. It's really the same types of research and process engineering.

And the Critical Materials Institute is making progress on more efficient methods for separating rare earth elements from one another, from recovering lithium from domestic brines, to recycling rare earth magnets from hard disk drives, for example.

CHAIRMAN. I want to have Mr. MacGillivray speak specifically to the process there at Ucore, but first, Mr. Barrios, I understand that you are working on a project with DOE's Critical Materials Institute to improve recovery rates for minerals. What can you describe about this partnership that you are working on with the Critical Materials Institute?

Mr. BARRIOS. In our copper deposit in Utah, copper is a gateway material. In addition to copper, we produce olibanum, gold, and silver. But we also extract other metals like rhenium, which is quite critical to the U.S. national defense, and it's one of the critical materials.

And what we've been doing in this work with the Department of Energy and the Critical Materials Institute is to continue exploring how we can extract more rhenium, but also look at other potential metals that we could extract together with copper. One of them is tellurium, which is used to increase efficiency in solar, converting solar into electricity and it increases the efficiency by about ten percent, exacting a key contributor to the challenges of climate change. So we are working very actively now to try and understand what other minerals we can actually produce at our Kennecott Copper Mine.

CHAIRMAN. Good, good.

Mr. MacGillivray, I want to have you go into a little more detail for the Committee about this MRT technology, the Molecular Recognition Technology, because as you have described, this technology, I do not know whether we describe breakthrough as the appropriate term, but if it is a reality that the permitting for this chemical extraction process is not going to be allowed in this country, then much of what we are talking about becomes moot and we just say we will rely on it for others.

But I have had the benefit of a brief from Ucore on the specific technology. If you can, in layman's terms for the Committee's benefit, please explain what MRT actually does, how it is different from the chemical extraction process, and really why it works environmentally.

Mr. MacGillivray. Thank you for the question.

So I mentioned earlier that Ucore recognized that solvent extraction would not be permittable in the United States, let alone Southeast Alaska, which is very environmentally sensitive. And when we shopped worldwide, we landed in Utah.

There's a company there by the name of IBC Advanced Technologies, and they have been in the metals separation business for over 20 years. They use a ligand technology. This is a highly selective, kinetically rapid, selective method of absorbing individual elements onto their ligand. They had not developed ligands for rare earth elements. They were working with other metals commercially. So Ucore invested money with them to specifically develop a ligand specific to rare earth elements. They conducted bench scale testing using the Bokan-Dotson Ridge ore and individually separated all 15 of the lanthanides that we had for that project. Since that time, we've invested into a pilot plant and up scaled that technology into, you know, a pilot plant scale.

The next step that we would like to pursue is the commercialization of this technology. We're very confident that it will work and be able to supply rare, individual rare earth elements for the United States, whether those sources come from recycling or heavy minerals sand by-product or ore itself from the Dotson Ridge project.

But I think initially we're going to concentrate on by-products where we can find concentrates of rare earths, like Dr. Eggert mentioned, and then using that clean, green technology, be able to permit a facility in the United States.

Chairman. If you have gotten to the point where you believe the pilot project is successful, why do you feel that you need federal resources to assist with commercialization?

Mr. MacGillivray. Well naturally, we would probably start out fairly small, and we are competing in a monopoly situation with the Chinese. So the private sector markets are somewhat supportive but it's really taking that first leap, that little shot, as to what kind of advantage can we provide to get that first step. And we believe that domestic supply of individual rare earth elements in the United States is the necessary first step.

Chairman. Has there been any interest expressed by the Department of Defense?

We have talked a lot about security here, security of supply and all that entails and specific as to China and China's role when we are talking more about rare earths. Have you had any expressed interest from DOD?

Mr. MacGillivray. Certainly some of our experienced consultants here in Washington, have a history with the Department of Defense, so we're very strong in those communications in that area.

Again, the crux of the entire situation is the viability of the technology to actually separate out these. So we need to have, sort of, a commercial scale plant to initiate that supply to build that confidence and then the things start rolling.

Chairman. And then to go to a question that was raised by Senator King and the requirements for being able to do the separation at the site.

You mentioned the location of the project that Ucore is looking at—Bokan—is in Southeastern Alaska.

Mr. MACGILLIVRAY. Yes.

CHAIRMAN. My hometown, where I was born and raised in that same region, is environmentally sensitive.

If you were to go to commercialization, what assurance can you give me to provide to Alaskans that there is a level of environmental safety and attention to the nature of the environment there and that it would not be at risk?

Mr. MACGILLIVRAY. Certainly.

I believe that Alaskans have great confidence in the scrutiny that the state provides during the permitting process. State engineers would certainly take a look at our technology, understand the water balance that goes on within the processing and the chemical characterization each step of the way.

The Bokan-Dotson Ridge Project is actually fairly innovative in itself in that should we be able to get that mine up and running, that due to x-ray ore sorting and MRT technology, we'll be backfilling 100 percent of the milled tailings back underground. So the project description for that project, the mine project, is very, you know, it's something to be proud of.

With respect, now we look more toward, well, the first step, because we like to phase our approach to entering into the rare-earth space. The first step would be building the separation plant in isolation. So probably not on the project site, but in a good location with infrastructure.

All I can say is that the permitting regime is strict and the reviews will be thorough, and I believe that once understood this ligand-based technology is exemplary.

CHAIRMAN. It always gets your attention when a process that involves issues related to toxins, to toxic waste here, can be referred to as a "green" technology. So there is a lot of interest in what you are pursuing.

Mr. MACGILLIVRAY. Right.

CHAIRMAN. I look forward to talking with you more about it.

Let me turn to Senator Cortez Masto.

Senator CORTEZ MASTO. Thank you, Madam Chair, and I apologize for having to leave. I have a competing Banking Committee meeting going on, but I am very interested in the discussion today.

Dr. Eggert, you may have talked a little bit about this while I was gone, and this is the issue of lithium mining.

In Nevada, we have lithium mining and it is important to both a booming technology industry—we have Tesla there as well as our geothermal companies that procure an abundant amount of geothermal resources in my state.

I am curious. Are there technologies that help both these industries utilize that lithium so that it is compatible and they are not necessarily competing against one another?

Dr. EGGERT. Well, within the research consortium that I'm involved in, the Critical Materials Institute, we are working on processes to recover lithium from geothermal brines in the salt and sea area and then process it into a form and a purity that allows it to be used in lithium ion batteries. We have had some technological success and the work that we are now working on with an industry partner is scaling that, proving that, at larger than in a test tube or a bench top scale but also at a larger scale as well.

Senator CORTEZ MASTO. Thank you. I appreciate that.

Thank you, Madam Chair, I appreciate the opportunity to have a second round of questions.

CHAIRMAN. Thank you.

Gentlemen, this has been, I think, very instructive, good information. We have had a couple different hearings in the Energy Committee over the past several weeks focused on infrastructure, and those hearings will continue. Senator Stabenow raised the question of infrastructure in view of what we are talking about with gaining access to critical minerals and our resources. In every infrastructure hearing that we are talking about, it comes down to permitting and a level of certainty. It is clear to me that we have much work that can be done in those spaces.

I think we heard today that there are opportunities to do a little bit better, whether it is designation of a lead agency or firm deadlines, but all that we can be doing from the perspective of providing some level of certainty to those who are looking to take the risk.

We have not talked about the risk that is inherent in the commodities market, that prices go up and prices go down. I can recall several decades ago being at the ribbon cutting at a Molybdenum—it is so hard to say, that is why we say Moly—at a Moly mine outside of Ketchikan. I was there for the ribbon cutting, and that was it. That was all she wrote. The price of Moly went down, and I do not believe there was ever any resource that was extracted from that mining venture.

That is a risk that is inherent within the industry. I think, Senator Cortez Masto, coming from a mining state, that there are years when the state's economy is good and strong and robust and others when it is not so much. So much of it is pricing beyond our control.

But those things that we can control, it seems to me, we should make a better effort to, again, provide for some level of certainty and a process that is fair and reliable.

I appreciate, Mr. MacGillivray, you saying that the permitting in these areas needs to be rigorous. We want to ensure that we are meeting good, strong, environmental standards so that the land that we are charged with taking care of is respected.

But there is a balance here that at some point you say, when you have overlaying bureaucracies, when you have overlaying or perhaps inconsistent regulation that causes confusion, that that adds to costs because you have duplication of effort. There is a rationale for streamlining, but streamlining does not necessarily mean environmental shortcuts.

How we lay that all down, how we make it work so that industry can operate is what, I think, we need to be doing. We do not want to be the country with a bad environmental track record. We will not accept that. But we also want to be the country that has greater predictability so that investors can look at the United States with, perhaps, a little more enthusiasm than we might have seen.

So we have opportunities with the resources. We thank the people at USGS for the good work that they do.

I would certainly encourage us, and I will be looking to the budget as the Chairman of the Interior Subcommittee that has the oversight of the USGS budget, I would like to see us making sure that the efforts to do better by our resources and understanding our resources are maintained.

But those of you that are in the industry, those of you that are helping to educate those who become part of the industry, know that we appreciate the contributions that you bring to the table.

For those who have joined us from outside the United States, again, we welcome your contributions and all that you have provided here today.

With that, we stand adjourned and thank you very much.

[Whereupon, at 11:50 a.m. the hearing was adjourned.]

APPENDIX MATERIAL SUBMITTED

Committee on Energy and Natural Resources
March 28, 2017 Hearing on Critical Minerals
Questions for Dr. Murray Hitzman

Questions from Chairman Murkowski

Question 1: Do you agree that the United States' dependence on foreign sources of minerals is problematic, and presents a strategic vulnerability for us? Can you each tick through some of the threats this presents for us, whether to our economy or our security?

Response: U.S. reliance on foreign sources of mineral raw materials for which production is highly concentrated in a single country, countries with high governance risk, or both, could constitute a significant risk to our economic and national security interests. The U.S. is 100 percent dependent upon foreign countries for 20 minerals. Examples include the rare earth elements, gallium, graphite, indium, manganese, niobium, and tantalum. Additionally, the U.S. imports more than 50 percent of our supply of 30 minerals such as germanium, rhenium, cobalt, lithium, and platinum, among others.[1] Combined, these 50 minerals have uses ranging from everyday commodities to smartphones to weapons systems.

Question 2: What percentage of the USGS' budget goes to minerals work each year? What is the historical trajectory for that work? Do you believe that funding for minerals work should increase, at least until our dependence on foreign minerals begins to decrease?

Response: In FY 2001, the Minerals Resources Program (MRP) was appropriated $54.5M, which represented 6.17 percent of the overall USGS budget. Since that time, MRP's appropriation as a percentage of the bureau's total appropriation has declined steadily. The graph below shows the historical trend in MRP's percentage of total USGS appropriations since FY 2001, culminating in an FY 2017 President's Budget request of $48.7M, or 4.17 percent of the total proposed USGS appropriation.

---

[1] Mineral Commodity Summaries 2017, pages 6-7.

Committee on Energy and Natural Resources
March 28, 2017 Hearing on Critical Minerals
Questions for Dr. Murray Hitzman

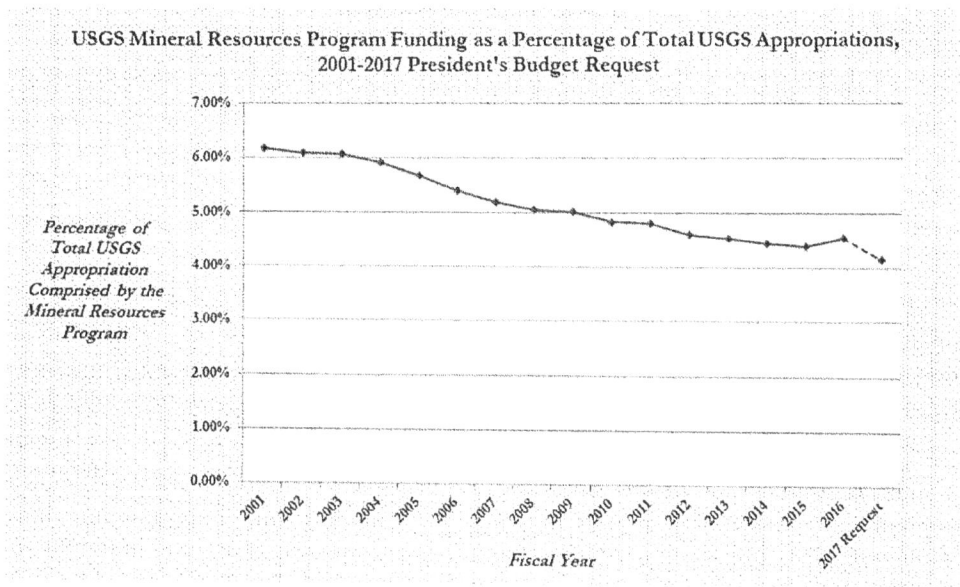

Question 3: The USGS was established to survey and classify the mineral reserves in the United States. The Survey went through a re-organization in 2013 which demoted the minerals program to just one part of one of its seven mission areas. Dr. Hitzman, you state in your testimony that "the USGS, stands ready to fulfill its role as the federal provider of unbiased research on known mineral resources [and] assessments of undiscovered mineral resources." As the head of the energy and minerals mission area, what is USGS doing to fulfill that commitment and to re-prioritize its minerals mandate?

Response: The USGS Minerals Resources Program continues to investigate the geology of known mineral deposits and utilize these data, along with historical data and new geological data produced by geological mapping and geophysical surveys, to produce mineral assessments of undiscovered mineral resources as requested by agencies within the Department of Interior and as directed by Congress. The program recently released the report "GIS-based identification of areas that have resource potential for critical minerals in six selected groups of deposit types in Alaska"* and is soon to release a report entitled "Critical mineral resources of the United States." In addition to mineral resource assessments, the USGS continues to collect information on the production, consumption, and recycling of mineral resources in the US and globally. This includes the annual Mineral Commodity Summaries report which identifies the import reliance of the US for many important and critical mineral commodities. The effort has been expanded to include development of a "criticality screening tool" that was featured in a 2016 report submitted to Congress by the Interagency Subcommittee on Critical and Strategic Mineral Supply Chains

Committee on Energy and Natural Resources
March 28, 2017 Hearing on Critical Minerals
Questions for Dr. Murray Hitzman

entitled, *Assessment of Critical Minerals: Screening Methodology and Initial Application.*

*Karl, S.M., Jones, J.V., III, and Hayes, T.S., eds., 2016, GIS-based identification of areas that have resource potential for critical minerals in six selected groups of deposit types in Alaska: U.S. Geological Survey Open-File Report 2016–1191, 99 p., 5 appendixes, 12 plates, scale 1:10,500,000, http://dx.doi.org/10.3133/ofr20161191.

Question 4: You mentioned in your testimony that we have not geologically or geophysically mapped the entire United States. Why is that data important, how is it used, and what do we have left to do?

Response: The USGS has completed detailed geological mapping and detailed aeromagnetic and radiometric surveying of approximately one-third of the United States. The data from these products are important in identifying geological areas that may be favorable for mineral deposits. Such data are utilized by the private sector to select regions for mineral exploration. Detailed geological mapping contributes to the discovery of new mineral commodities, informs responsible management of our mineral resources, and has the potential to decrease our reliance on foreign sources for raw processed mineral materials.

Committee on Energy and Natural Resources
March 28, 2017 Hearing on Critical Minerals
Questions for Dr. Murray Hitzman

Questions from Senator Hoeven

Questions: Lithium is one critical mineral which is used in energy storage, polymers, lubricants, pharmaceuticals, agricultural products, ceramics, and construction. In addition, the U.S. is the largest consumer of lithium metal for primary batteries and lithium aluminum alloys. However, our country's domestic lithium metal production was reduced by 50 percent last year and now represents only 10 percent of worldwide lithium production.

- Should we be concerned about this significant reduction in U.S. lithium metal production, requiring dependence on China and Russia for our lithium supply?

Response: Security of supply lies in diversity of supply. The global production of lithium, like many other mineral raw materials, is highly concentrated in one or two countries. Although it is found domestically here in the United States, it is increasingly not produced here.

- What policies should Congress consider to address this issue?

Response: As a science agency, USGS focuses on research and data on critical minerals and leaves policy and management decisions to other authorities. Congress has addressed this issue in the past. Title III of the Defense Production Act (DPA) (Pub. L. 81-774) provides the President broad authority to ensure the timely availability of essential domestic industrial resources to support national defense and homeland security requirements, by authorizing economic incentives to create, expand, and modernize production capacity.

Committee on Energy and Natural Resources
March 28, 2017 Hearing on Critical Minerals
Questions for Dr. Murray Hitzman

Questions from Senator Cortez Masto

Question 1: The USGS is working on the Mineral Database Deposit Project, which will be a database of all mines and mineral deposits in the U.S. What is the status of this project? Will technologies improve the effectiveness of the database because mineral deposits are more easily identified?

Response: The USGS is actively working on the Mineral Database Deposit Project (called USMIN) that collects existing information about mines and mineral deposits of the U.S. in a form that is readily accessible and in a format that will be directly useable by other Federal Agencies such as the Bureau of Land Management (BLM). The database is improved by technologies such as geographic information systems (GIS) that allow easy display of the information in a searchable and customizable format.

Question 2: Do you think technologies could improve mapping for potential mineral development? Could those technologies also work to protect more environmentally sensitive lands that support the outdoor recreation industry, communities, and wildlife?

Response: Mapping is an important function of the USGS and is essential for identifying areas of potential mineral development and informing responsible resource management. Technology continues to improve to provide better mapping and understanding of geological resources. The ability to date rocks using various isotopic techniques has revolutionized geologic mapping compared to the early days of the USGS. New developments continue on an annual basis such that it is now possible to date and distinguish rock units, and thus map them, with unprecedented precision. In addition, continually evolving geophysical technologies allow better characterization of the subsurface of the Earth and construction of much more accurate bedrock geological maps of areas where rocks are covered by thick soils or other overburden. This information contributes directly to effective understanding and management of environmentally sensitive lands that support the outdoor recreation industry, communities, and wildlife.

Question 3: Are there ways in which USGS, Department of Energy, and Department of Defense can work together more effectively to increase our domestic supply of critical minerals?

Response: Under the auspices of the Interagency Subcommittee on Critical and Strategic Mineral Supply Chains, the USGS and the Department of Energy have worked jointly to develop a "criticality screening tool" that is a method to quantify early warning criticality indicators across minerals. The Department of Energy, through the Critical Materials Institute, is focusing on the development of efficient extraction and separation technologies to help maximize the recovery of several commodities, notably rare earth elements, deemed critical. The Department of Defense Logistics Agency is utilizing the "criticality screening tool" at this time.

Question for the Record from Senator Cantwell
Senate Committee on Energy and Natural Resources

Oversight hearing on the United States' increasing dependence on foreign sources of minerals and opportunities to rebuild and improve the supply chain in the United States

March 28, 2017

For Dr. Hitzman:
In response to a question from Senator King about whether mining companies must pay federal royalties on locatable minerals on federal lands (see the transcript below), you answered that they do. In fact, under the Mining Law of 1872 that continues to govern the extraction of hardrock minerals on federal lands, no federal royalty is collected. Could you please correct the record and clarify your answer for the committee?

"Senator King: I want to talk a bit about permitting, and I know I am running out of time. Quick question. If federal lands are used for one of these mines, are there royalties paid to the taxpayers for the extraction? Mr. Hitzman?

"Dr. Hitzman: Yes.

"Senator King: Okay, so there are royalties that come back for whatever the value is of the mined minerals."

Response: Thank you for the opportunity to clarify the response for the record. There is no royalty collected from mineral development under the Mining Law of 1872, 30 U.S.C. §§ 22 et seq., as amended. Minerals on federal lands that are mined under the Mining Law include both metallic minerals (gold, silver, lead, copper, zinc, nickel, etc.) and nonmetallic minerals (fluorspar, mica, certain limestones and gypsum, tantalum, heavy minerals in placer form, precious gemstones), and certain uncommon variety mineral materials. While the Federal government does not collect royalties under the hardrock mining program, it does receive revenue from an annual "rental" of $7.75 per acre for the use of mining claims. In addition, the operator must post a financial guarantee sufficient to cover reclamation costs. Lastly, rental fees net the Federal government approximately $63 million, of which about $39 million is used to fund the management of the program.

U.S. Senate Committee on Energy and Natural Resources
March 28, 2017 Hearing: The United States' Increasing Dependence on Foreign Sources of Minerals and Opportunities to Rebuild and Improve the Supply Chain in the United States
Questions for the Record Submitted to Mr. Alf Barrios

Questions from Chairman Lisa Murkowski

Question 1: *Do you agree that the United States' dependence on foreign sources of minerals is problematic, and presents a strategic vulnerability for us? Can you each tick through some of the threats this presents for us, whether to our economy or our security?*

As outlined in the USGS 2017 Mineral Commodities Summaries, the US is becoming increasingly reliant on foreign minerals.

In the case of Rio Tinto, our copper production provides us opportunities to capture co-product metals and minerals -- e.g. rhenium, selenium, tellurium, cobalt, the rare earths, and gallium -- that, according to USGS data, show very high US import dependencies and are often considered critical and strategic to national defense. It is important, as we see it, for US policy to take into account that primary industrial metals are effectively our gateways to metals and minerals increasingly key to high-tech, alternative energy and national security applications.

However, as stated in my oral testimony, enhancing the US' ability to access its own resources does not mean we should raise barriers to imported materials. Nowhere are the mutual benefits of trade more apparent than the integrated supply chains in North America, for example in aluminum, where inputs from Canada make US manufacturers more competitive and vice versa.

Question 2: *If you had certainty that your permitting would proceed according to a clear and predictable 2-3 year schedule, what level of investment would Rio Tinto be looking to make in the United States? Or put another way, how much is Rio Tinto looking to invest with a project like Resolution Cooper? Are you asking for any federal support, or just timely federal decisions? How much private investment in the United States economy comes along with a project of that scale?*

Rio Tinto is proud of its 100-plus years of operations in the United States. As the Committee is aware, mining operations require significant capital expenditure throughout the life cycle of an operation. For example, we have invested over $1.3 billion in our Resolution Copper project (Superior, Arizona) to date. The project will require hundreds of millions of additional investment during the permitting stage of the project. Construction required prior to production of copper in commercial quantities will be billions more.

The Resolution Copper Mine is expected to have an economic value of several billion over the estimated life of the project and create several thousand direct and indirect jobs.

Behr Dolbear routinely publishes a Ranking of Countries for Mining Investment. In their 2014 report the United States ranks in the top 3 countries for mining investment, but the report makes one noteworthy exception in the category of permitting delays where *"Permitting delays are the most significant risk to mining projects in the United States."* - Rio Tinto is one of, if not the

U.S. Senate Committee on Energy and Natural Resources
March 28, 2017 Hearing: The United States' Increasing Dependence on Foreign Sources of Minerals and Opportunities to Rebuild and Improve the Supply Chain in the United States
Questions for the Record Submitted to Mr. Alf Barrios

largest, investors in exploration across North America. We believe despite data indicating a decreased rate of exploration spend, that the US would benefit from increased investment if there was more certainty associated with permitting timeframes.

Questions from Senator Catherine Cortez Masto

Question 1: *Do you believe that chronically underfunded and understaffed agencies affects permitting deadlines?*

Yes, in our experience, permitting delays can be attributed to a number of factors, including understaffed agencies and lack of funding to develop a qualified staff experienced in the permitting process.

Question 2: *According to a GAO report, **BLM and Forest Service have tried to set up pre-mine plans to gather more information up and consult with companies before the official review process.** Do you believe meetings like these are helpful in trying to expedite permitting delays?*

Yes, we embrace and promote BLM and Forest Service pre-NEPA plans as well as proactive discussions and regular communication between the private sector and government before and during NEPA. This approach is helpful.

Another good approach is consistent application of the Council on Environmental Quality's (CEQ's) NEPA regulation (sec. 1501.8) that *"the agency shall set time limits if an applicant for a proposed action requests them."* Many applicants may be reluctant to make such a request, but a reasonable and agreed upon schedule serves as a good reminder to the agency the importance of moving the process along. It also helps bring clarity to the roles and responsibilities of the applicant and the agency as well as permitting timeframes which is important from an investment standpoint.

Question 3: Do think the Administration's proposed budget cuts will affect the Critical Minerals Institute's ability partner with you to promote these types of technologies?

At Rio Tinto's Garfield copper smelter in Utah, we are partnering with the U.S. Department of Energy's Critical Materials Institute (CMI) to find new ways to fully recover and recycle the minerals that future technologies will require. This means not just looking at more efficient ways to process and extract minerals from the ground, but also "urban mining" of electronic waste.

At this stage we do not believe our partnership with CMI will be impacted by proposed budget cuts to DOE, however, we believe funding should be protected for these valuable programs.

United States Senate Committee on Energy and Natural Resources
March 28, 2017 Hearing: The United States' Increasing Dependence on Foreign Sources of Minerals and Opportunities to Rebuild and Improve the Supply Chain in the United States
Questions for the Record Submitted to Dr. Chris Hinde

Questions from Chairman Lisa Murkowski

Question 1: Do you agree that the United States' dependence on foreign sources of minerals is problematic, and presents a strategic vulnerability for us? Can you each tick through some of the threats this presents for us, whether to our economy or our security?

Response: Yes, I agree, the United States' dependence on foreign sources of minerals is a potential threat to the continued growth of the local economy, and this reliance on others also increases the country's vulnerability to hostile trade or political action.

The greatest threat is from unforeseen restrictions in the supply of metals or minerals that are important ingredients in manufacturing processes, especially for those metals where there is no obvious substitute. Critical metals with an import dependency of over 50% include cobalt, gallium, indium, platinum group metals, rare earths and tellurium. The rare earth elements are particularly problematic, being critical in the manufacture of catalysts and many alloys. The United States achieves only 4% of global production despite having an estimated 9% of the world's REE reserves, and is 70% reliant on foreign supplies as a result.

Question 2: Are you able to provide an estimate of how much money the United States is losing out on in terms of potential investment due to our slow permitting process for new mines?

Response: It is difficult to evaluate empirically the loss of investment caused by slow permitting without a separate study. However, comparison can be made with exploration expenditure in Australia and Canada, which have similar mineral wealth and environmental standards to those prevailing in the United States. In 2016, the United States saw only US$500 million spent on the search for metals (other than iron ore), whereas Australia benefitted from expenditure of US$897 million and Canada of US$971 million. This shortfall in exploration expenditure can be expected to lead eventually to lower metals production in the United States as the ore reserves of existing mines become exhausted, and will lead to an increased import dependence for local manufacturers.

These differential exploration expenditures are even starker when population and economy sizes are taken into account. The United States' population is almost nine times that of Canada and over 13 times that of Australia. The country's economy (measured as GDP in 2016) is over 13 times that of Australia and almost 17 times that of Canada. As a result, the latter country receives 32 times more in equivalent exploration expenditure than the United States in terms of national GDP, and Australian exploration is equivalent to a per capita US$37, compared with only US$1.60/person in the United States.

United States Senate Committee on Energy and Natural Resources
March 28, 2017 Hearing: The United States' Increasing Dependence on Foreign Sources of Minerals and Opportunities to Rebuild and Improve the Supply Chain in the United States
Questions for the Record Submitted to Dr. Chris Hinde

Questions from Senator Al Franken

Question 1: Mining is a way of life in northern Minnesota. My state is home to the great Iron Range where the vast majority of iron ore used in our country's steel mills is produced.

The Iron Range has a number of different resources, including a substantial copper-nickel deposit. And there's currently an ongoing debate in Minnesota about if and how to use these resources. This is especially pertinent because the economic potential and job creation is alluring. Now, northern Minnesota is also home of the pristine Boundary Waters Canoe Area Wilderness—the most visited wilderness area in the country—and Voyageurs National Park. So I believe that any decisions must follow the process and must be firmly based in science.

We also have large taconite, or low-grade iron ore, deposits and over the last few years, the Iron Range has suffered greatly as global steel overcapacity has put U.S. mills out of work and left our mines with nowhere to ship their product. Last year, the Obama Administration levied heavy tariffs on Chinese steel imports to start to remedy this problem, and although some of our miners are now getting back to work, there are still as many as 500 miners still out of work. China obviously has been producing far more steel than their domestic industry has demanded recently. Do you expect their domestic demand for steel to increase or will they continue to have significant overcapacity in the near term?

Response: China has been consuming far more iron ore and steel on both a GDP and per capita basis than is historically typical for other nations. Most of this demand can be attributed to the country's aggressive development of infrastructure. Consumption of iron ore and steel seems unlikely to remain at these current very high levels so, unless there is an unexpected and dramatic reduction in the local production of these metals, the exports of iron ore and steel will remain high.

However, it should be noted that the Chinese government has announced measures to tackle overcapacity in its steel industry, partly driven by environmental pollution concerns, with a targeted closure of 100-150 million tonnes by 2020. It does appear that the government has recently been intensifying and re-enforcing its crackdown as its policy has met a number of challenges. These include some already idled, and not necessarily producing capacity, being reported as curtailments; restarts of some idled capacity (as prices increased) and the addition of new production capacity. Furthermore, some regional governments may not have been effectively complying with the central directives.

Question 2: Does China's steel industry appear to follow market-driven principles? Or is it reliant upon Chinese government subsidies?

Response: Most Chinese industry does not operate purely on the basis of free, capitalist, market principles. Market forces are important, but a number of sectors, including iron ore and steel, are heavily influenced by socio-political concerns and by central government direction. Because of

United States Senate Committee on Energy and Natural Resources
March 28, 2017 Hearing: The United States' Increasing Dependence on Foreign Sources of Minerals and Opportunities to Rebuild and Improve the Supply Chain in the United States
Questions for the Record Submitted to Dr. Chris Hinde

the political framework, the national and regional governments in China are able to take longer-term decisions than are available to western governments. This means that metals production can be dictated by strategic imperatives, such as job creation, supply-chain security, or even environmental concerns, rather than corporate profits.

Questions from Senator Joe Manchin III

Question 1: In your written testimony, you focused on the importance of "rigorous permitting" to ensure our nation's potential mines are permitted within a timeframe that does not lead to delays and major losses in value. In 2014, I introduced the Regulatory Fairness Act which would have prevented the EPA from retroactively or preemptively vetoing water permits associated with mines. This happened in West Virginia with a Minho Logan mine. And it's my understanding that Pebble Mine in Bristol Bay, Alaska is still held up in litigation following a preemptive veto.

Would the Regulatory Fairness Act help provide mine developers with the certainty they need?

What else can be done specifically?

Response: Yes, it would seem that the Regulatory Fairness Act would contribute to the certainty of mine-development decisions. As noted in my testimony, mining companies seek, beyond anything else, a process that is certain, in terms of both the consistency of the requirements and of the timetable. A specific improvement (for all parties) would be a consolidation of the requirements so that repetitive and overlapping tasks are minimized.

Question 2: Resource extraction in the United States has become increasingly politicized in recent years. Dr. Hinde, in your written testimony you discuss an independent study that S&P conducted to analyze the impacts of permitting delays. The report showed that project risks increase when permits are delayed, leading to a decrease in value of the project. I understand the importance of providing certainty to operators seeking undertaking mining projects. That said, as a former Governor and a Senator, I also believe it is important that landowners and the public are treated fairly and given the opportunity to be heard.

What do you believe can be done to streamline the permitting process and remove duplications while still ensuring robust public engagement?

In your written testimony you also discuss S&P research that is showing that manufacturing activity is returning to the USA. This is driven by manufacturers desire to mitigate risks to their supply chain, as well as consumers concerns for corporate accountability. What can Congress do to ensure that this encouraging trend continues?

United States Senate Committee on Energy and Natural Resources
March 28, 2017 Hearing: The United States' Increasing Dependence on Foreign Sources of Minerals and Opportunities to Rebuild and Improve the Supply Chain in the United States
Questions for the Record Submitted to Dr. Chris Hinde

Response: Without any loss of environmental standards or investigative rigor, the permitting processes in both Canada and Australia are much more efficient than in the United States. This is primarily for two reasons. First, one agency acts as a lead in the process, and ensures that requirements are consolidated, and the necessary tests and checks arranged in an orderly fashion. Second, the environmental assessments are organized by the company itself, using reputable third-party consultancies, which are then subject to scrutiny, rather than left to various agencies and stakeholders to manage.

Yes, manufacturing does appear to be returning to the United States. Congress could enhance this trend by taking measures to reduce risks to the supply chain and by encouraging public awareness of corporate responsibility. The first of these measures could be facilitated by boosting the local supply of metals through a more efficient permitting process (as outlined in the previous paragraph). The second measure would benefit from a program of making the public aware of the social, economic and environmental benefits of consuming local materials.

Questions from Senator John Hoeven

Questions: In your testimony, you said mining is a business where certainty is important, especially for long-term, capital-intensive projects. You also stated that, "New mines can typically lose over one-third of their economic value as a result of even relatively small delays in reaching production."

You mentioned that it takes on average seven to ten years to secure the permits needed for mines to reach production in the United States. But in other countries – like Canada and Australia – with comparable mining resources equally stringent environmental regulations, they have average permitting periods of two years.

- Why are there significant differences between the United States and these other countries with shorter permitting timelines?

- What are these countries doing – on the legal, tax, or regulatory fronts – that the United States is not?

- What are some things that U.S. government can be doing to ensure that the permitting process is fair, certain, and efficient?

- What policies should Congress consider in order to improve this process?

Response: According to our analysis, the time for a mine to reach commercial production in the United States is typically between two and four times the equivalent time taken in Australia and Canada, which have similar environmental standards. In both these countries, one agency acts as a lead in the process, and ensures that the requirements of multiple stakeholders are consolidated,

United States Senate Committee on Energy and Natural Resources
March 28, 2017 Hearing: The United States' Increasing Dependence on Foreign Sources of Minerals and Opportunities to Rebuild and Improve the Supply Chain in the United States
Questions for the Record Submitted to Dr. Chris Hinde

with the necessary tests and checks arranged in an orderly fashion. Also, the environmental measurements and assessments are organized by the mining company, and then subject to scrutiny, rather than left to the various agencies, which may have overlapping requirements.

These other countries are doing relatively little different to the United States in terms of their overall legal, tax and regulatory arrangements, it is rather just the efficiency in which they apply similar rules. Everyone involved in these permitting issues should be seeking a fair, certain and efficient process, and certainly there should be no weakening of the environmental protection. Congress simply needs to ensure that all of the various imperatives in the process are clearly documented at the start, and that a proper implantation plan, with defined timeline, is established.

United States Senate Committee on Energy and Natural Resources
March 28, 2017 Hearing: The United States' Increasing Dependence on Foreign Sources of Minerals and Opportunities to Rebuild and Improve the Supply Chain in the United States
Questions for the Record Submitted to Mr. Randy MacGillivray

Questions from Chairman Lisa Murkowski

Question 1: Do you agree that the United States' dependence on foreign sources of minerals is problematic, and presents a strategic vulnerability for us? Can you each tick through some of the threats this presents for us, whether to our economy or our security?

Response 1: Ucore strongly believes that the United States' dependence on foreign source of minerals and materials is both problematic and introduces a dangerous risk to our domestic economy and our national security.

From a defense perspective, numerous weapons systems and platforms are reliant upon certain materials to function properly. These materials, such as rare earth elements, are used in a variety of applications from making strong magnets used in small actuators to control the flight trajectory of smart bombs to thermal barrier coatings in high temperature sections of jet aircraft engines.

At present, the People's Republic of China controls the production of rare earths elements producing close to 95 percent of the world's output. An embrace of environmentally destructive mining and processing techniques has allowed the Chinese to corner the market for rare earths, set global prices, force competition from the market, and effectively control entire supply chains by maintaining a strong monopoly. We have previously witnessed China exercise its control on the supply chain when, after a dispute with the nation of Japan in 2010, which is threatening again in 2017, the Chinese introduced quotas on the export of rare earths, drastically reducing exports which sent prices soaring. While this scenario enticed many companies to enter the market due to the lucrative prices in the wake of the Chinese export restriction, a subsequent World Trade Organization determination ruled against China and forced the Chinese to abandon the quotas and flood the market with material. Prices for rare earth elements plummeted and many businesses were forced from the market as their business cases were no longer profitable.

Today, the United States lacks even a single operating producer of rare earth elements. The last remaining major producer of rare earths in the United States filed for bankruptcy in 2015, a casualty of Chinese trade policies.

Should the Chinese ever curtail U.S. access to these essential materials, the U.S. would be woefully unprepared to meet domestic commercial and military demand. The United States must compete with the Chinese in this market sector before it is too late. A deficient supply of critical and strategic rare earth metals and the associated manufactured products produced from these metals would be a detriment to the ability of suppliers in the renewable energy and automotive sectors and the U.S. military and its suppliers to provide critical high tech engineered products.

Question 2: Others are paying attention to this issue, especially China, but the United States does not seem to be concerned with our increasing dependence on foreign mineral sources. What do we need to do to make these issues relevant to U.S. policymakers?

United States Senate Committee on Energy and Natural Resources
March 28, 2017 Hearing: The United States' Increasing Dependence on Foreign Sources of Minerals and Opportunities to Rebuild and Improve the Supply Chain in the United States
Questions for the Record Submitted to Mr. Randy MacGillivray

Response 2: It would not be prudent to allow this issue to linger until the next crisis hits. Continued political pressure through the introduction of legislation and the accompanying education through debate are necessary to stimulate the actions required to provide competition in the marketplace and a domestic supply of critical and strategic minerals. Ucore appreciates the legislative attempts made during the last Congress including S.3203 which provided support for the commercialization of a clean green separation technology.

Additionally, Ucore would like to recognize the efforts being made in the House of Representatives this year with the introduction of H.R. 1407, the Materials Essential To American Leadership and Security Act ("METALS Act"). This bill would provide the necessary access to capital domestic rare earth technology companies need in order to finance their start-up ventures and bring new and innovative technologies from bench and pilot scales to full commercialization. Legislation designed to address the lack of access to capital in the wake of the highly publicized bankruptcy of the last major domestic producer of rare earths in 2015 must be a key component of any strategy designed to promote the development of the domestic industrial base for the production of rare earths.

U.S. policymakers must appreciate that developing an American based domestic commodity and product supply chain will take years of development and that market dominance by a foreign country may have dire effects in the interim period. Continued reliance on foreign sources of critical materials while failing to address the lack of a domestic industrial base for their production will only exacerbate the problem should a time ever arise when U.S. access to foreign supply is restricted.

Questions from Senator Joe Manchin III

Questions: West Virginia University participates in the U.S. Department of Energy National Energy Technology Laboratory's ongoing program to recover rare earth elements from coal and coal byproducts. Acid mine drainage from pre-law coal mines is a source of pollution in the Appalachian region. As it turns out, when these areas are treated to meet current regulatory requirements, the process produces solids enriched in critical rare earth elements. Right now, West Virginia University and its partners work to sample and analyze acid mine drainage solids from 120 acid mine drainage treatment sites at coal mines across the northern and central Appalachian coal basins in West Virginia, Pennsylvania and Ohio. They work together to develop a cost-effective process to treat and recover rare earth elements from the sludges produced at these mines.

Do you agree that this technology would help enhance U.S. supply of critical minerals, thereby better protecting us against supply disruptions?

United States Senate Committee on Energy and Natural Resources
March 28, 2017 Hearing: The United States' Increasing Dependence on Foreign Sources of Minerals and Opportunities to Rebuild and Improve the Supply Chain in the United States
Questions for the Record Submitted to Mr. Randy MacGillivray

Response: Processes that can economically concentrate rare earth elements to enriched concentrations that can then be prepared to separate into individual rare elements is a step toward domestic independence of these strategic and critical materials. From the description provided, molecular recognition technology ("MRT") would be a good candidate for testing the final separation of these rare earth concentrates as MRT uses an acid based pregnant leach solution for processing. The United States should pursue multiple avenues for developing rare earth feedstocks including both proven extraction methods, such as MRT, and new, innovative solutions.

Do you see potential here for commercialization of this process?

Response: The key aspect is the initial step of concentrating the sludge into concentrations that have a significant value per kilogram. In this way, the value of the contained metals will provide a commercialization strategy and carry the cost of overall processing and separating into individual oxides. Investment in the commercialization of this process should only be made after the economic viability of recovery from coal and coal by-products has been established and any investment should be made in tandem with other, proven approaches to ensure the U.S. promotes a diverse and robust domestic supply chain for rare earth elements.

Questions from Senator Bill Cassidy

Question 1: Regarding the supply of mineral imports, to what degree has Chinese mineral market manipulation fulfilled Chinese economic hegemony?

Response 1: Over the past decade, the Chinese have successfully consolidated the vast majority of the world's output of rare earth elements in China. An embrace of destructive environmental practices including outdated refining and separation technologies and a willingness to pollute the landscape with waste products have allowed the Chinese to undercut the prices of foreign producers and effectively manipulate the market for rare earths.

Market analysts suggest that China controls 95% of the world wide supply of individual rare earth elements including complete control of the production of rare earth metal from oxide. This monopoly affects the ability of secondary manufactures of magnets and other products with significant rare earth metals to secure a stable supply. This had led to a situation where the supply chain for rare earth magnets, crucial components of numerous defense applications including actuators used in flight control surfaces, runs through China. Despite proven reserves of rare earths in the U.S., the commercialization of proven technologies to separate the materials in an environmentally friendly and green fashion, and the commercialization of a technology to process the materials from oxides to metals to alloys, due to continued market manipulation on behalf of the Chinese, has prevented the establishment of a domestic industrial base for the production of rare earth metals and products. Furthermore, owing to the highly publicized bankruptcy of the last major domestic producer of rare earths, and after that company amassed

United States Senate Committee on Energy and Natural Resources
March 28, 2017 Hearing: The United States' Increasing Dependence on Foreign Sources of Minerals and Opportunities to Rebuild and Improve the Supply Chain in the United States
Questions for the Record Submitted to Mr. Randy MacGillivray

more than $1 billion in debt, the ability to access capital to fund new and innovative technologies has disappeared. This bankruptcy, caused in part by unfair Chinese trade practices, has left capital markets unwilling to fund domestic projects and further increased U.S. dependence on China for rare earths.

Question 2: To what degree can U.S. and North American mineral resource development supplant need for supply outside of America?

Response 2: The development of active American rare earth mineral deposits remains a critical first step toward ensuring that the U.S. develops the ability to procure rare earth material for defense applications should U.S. access to foreign sources of supply ever be restricted

The underlying issue regarding the potential for a domestic supply of rare earth materials is the critical separation stage, taking mined ore and producing an oxide which can then be made into rare earth metal. If a mine today produces a valuable rare earth concentrate, it must send that concentrate away for separation. Presently, China controls 100 percent of the world's rare earth metal production from oxide imposing a dangerous choke point in the supply chain for rare earths.

While a particular focus has been given to the potential recycling of existing rare earth products in the U.S., those materials must still be sent to China in order for separation of the individual rare earth elements to be completed and new products made to the specifications required. The issue is where, and at what cost, the concentrate from these mines would be separated into individual rare earth oxides to then be made into metal alloys. <u>Until proper investment is made into a domestic solution for the separation of rare earth elements, the U.S. will remain dependent on foreign supply chains for at least some, if not all, stages of rare earth production.</u>

Question 3: Aside from black sludge, what toxins would Chinese mining and processing release into the atmosphere and are in excess of what similar operations would be in the U.S.?

Response 3: The issue with the Chinese processing circuits is that they are using an outdated technique called 'solvent extraction'. This technique relies on the immiscibility of oil and water, and the selectivity of this method is very low, meaning that hundreds of cells and circuits are required to create a valuable product. The oil that is used in solvent extraction, once spent, reports to the tailings and creates a sludge and groundwater pollutant. We are not sure of issues related to releases to the atmosphere; whereas the potential to pollute groundwater is a significant environmental risk requiring the need for an environmentally sound separation technology to be applied in the U.S.

United States Senate Committee on Energy and Natural Resources
March 28, 2017 Hearing: The United States' Increasing Dependence on Foreign Sources of Minerals and Opportunities to Rebuild and Improve the Supply Chain in the United States
Questions for the Record Submitted to Mr. Randy MacGillivray

Question from Senator Catherine Cortez Masto

Question: You mentioned that Ucore has embraced the adoption of green technologies limiting the impact on the environment. How can the federal government improve R&D and commercialization of these green technologies and which programs are currently helping these efforts?

Response: Ucore has adopted molecular recognition technology ("MRT") and is now studying market entry to compete with foreign suppliers of rare earth elements in the U.S. The Chinese monopoly, however, has manipulated the market to the extent that it is only marginal for others to compete and very difficult for those looking to compete to attract private market investment. In order to be able to initiate a commercial venture, an economic incentive must be provided to take the venture from marginal to profitable either as an initial capital cost investment, or in a negotiated offtake agreement for domestic supply. Ucore has invested in, and completed the R&D portion of the pre-commercialization step, and is now poised to develop a small scale-commercial-demonstration plant to once again provide the U.S. with a domestic supply of rare earth materials

However, a lack of access to capital in the wake of the bankruptcy of the last major American producer of rare earths in 2015 has created a financing roadblock. Ucore has demonstrated its green separation technology on bench and pilot scales and is now seeking to commercialize MRT. Continued support for domestic solutions to the rare earth supply chain problem by the federal government is paramount to ensuring that the U.S. develops a domestic industrial base for rare earth production. Language introduced in the Senate during the last Congress and in the House of Representatives this year would begin to provide companies with a mechanism to fund proven technologies and domestic projects. Due to the nature of the use of rare earths in military and defense applications, the federal government must take steps to obviate our dependence on foreign sources of these critical and strategic materials. Congress should express its support of new, innovative, and green separation technologies developed in the United States and should assist companies seeking to commercialize these technologies and establish a domestic source for rare earth material.

U.S. Senate Committee on Energy and Natural Resources
March 28, 2017 Hearing: The United States' Increasing Dependence on Foreign Sources of Minerals and Opportunities to Rebuild and Improve the Supply Chain in the United States
Questions for the Record Submitted to Vice Admiral Kevin Cosgriff, USN (Retired)

Questions from Chairman Lisa Murkowski

Question 1: Do you agree that the United States' dependence on foreign sources of minerals is problematic, and presents a strategic vulnerability for us? Can you each tick through some of the threats this presents for us, whether to our economy or our security?

Answer: I agree that the U.S. dependence on foreign sources of many minerals and intermediate materials necessary for manufacturing presents challenges for a significant number of NEMA Member companies. Each company makes its own strategic sourcing decisions based on its understanding of risks and opportunities. Of course, large scale disruption of mineral supply chains, whether driven by government or business decisions or even by infrastructure failures such as highlighted by Sen. Stabenow during the hearing, would pose a material threat to manufacturers, with attendant economic implications for value chain partners and customers.

Question 2: Where do many of your member companies obtain the minerals needed to build their finished products? Would your members prefer to obtain raw minerals and materials from the United States, if that was possible?

Answer: Many of our Member companies source minerals from abroad, while many obtain minerals from domestic sources. As discussed, some mineral resources such as rare earth materials can at this point only be sourced from offshore. In order to maintain and improve their competitiveness, our members regularly evaluate their supply chains for risks and opportunities. If mineral inputs were available closer to home on competitive terms, each company would take that into account in their sourcing decisions.

Question 3: Others are paying attention to this issue, especially China, but the United States does not seem to be concerned with our increasing dependence on foreign mineral sources. What do we need to do to make these issues relevant to U.S. policymakers? And, how can we convince companies at the end of the supply chain to start talking about our foreign dependence and the significant problems that it creates?

Answer: These issues are relevant to U.S. policymakers, but their perceived importance relative to their other policy concerns is not as high. Public education efforts, such as the National Mining Association's "Minerals Make Life" campaign, would seem to be a valid approach to raising awareness. In general, manufacturers are reluctant to speak publicly and specifically about their vulnerabilities, especially in industries – such as those NEMA represents – where competition is intense.

Question from Senator Mazie K. Hirono

Question: Vice Admiral, in 2015, when you testified before this committee on this very topic, you explained that manufacturers are constantly trying to eliminate inefficiency in their

U.S. Senate Committee on Energy and Natural Resources
March 28, 2017 Hearing: The United States' Increasing Dependence on Foreign Sources of
Minerals and Opportunities to Rebuild and Improve the Supply Chain in the United States
Questions for the Record Submitted to Vice Admiral Kevin Cosgriff, USN (Retired)

processes and that waste is viewed as an inefficiency. You mentioned how DOE's Critical Materials Institute at the Ames Lab was focusing its research on how to better recycle critical minerals from products. Please comment on how you think the effort at Ames Labs is going so far? I would be interested in hearing the opinion of NEMA members of this research effort.

Answer: To date, a significant achievement of CMI research into reclamation and recycling critical minerals from end-of-life products has been to illuminate the difficulty thereof as well as the associated costs. More research is needed to develop lower-cost methods for extracting useful, useable, and beneficial materials, including design elements that could make a product to easier to deconstruct and recycle without materially reducing the product's utility, safety and performance during its service lifetime. Assessments must be based in science, using best-available data.

Questions from Senator Catherine Cortez Masto

Question 1: Rare earth elements are critical for innovation in clean energy, consumer electronics, computer applications, and health care technologies. These minerals are crucial for Nevada's clean tech sector which includes members of your association, employs an estimated 21,800 jobs, and is driving economic growth. Can the federal government, like with DOE's investment in the Critical Minerals Institute, help manufacturers achieve lower-cost operations and promote industry growth with technology breakthroughs?

Answer: The federal government should continue to support public-private collaboration on critical minerals research that would not take place if only private-sector resources were available.

Question 2: You have stated that USGS's work on The Minerals Commodity Summaries and other minerals information and analysis work is critical information that helps inform NEMA economic forecasters. Do you believe stable funding is crucial for this agency to do an effective job in helping private industry?

Answer: NEMA supports stable funding for the USGS Minerals Information Service.

Question 3: Do you think lithium-ion battery recycling will help manufacturers like Tesla to stabilize domestic supply in order to advance technologies that rely on lithium?

Answer: Lithium-ion battery collection and recycling is well underway in the U.S. See http://www.call2recycle.org/recycling-laws-by-state/#Nevada.

United States Senate Committee on Energy and Natural Resources
March 28, 2017 Hearing: The United States' Increasing Dependence on Foreign Sources of Minerals and Opportunities to Rebuild and Improve the Supply Chain in the United States
Questions for the Record Submitted to Dr. Roderick Eggert

Questions from Chairman Lisa Murkowski

Question 1: Do you agree that the United States' dependence on foreign sources of minerals is problematic, and presents a strategic vulnerability for us? Can you each tick through some of the threats this presents for us, whether to our economy or our security?

As I noted in my written testimony, it is not import dependence itself but rather risky import sources that are threats to U.S. users of mineral resources and the technologies that these resources underpin. Such is the case when imports come from one or a small number of production facilities, companies or countries – especially countries in which political decisions, restrictions on international trade, civil disruptions, or other developments may restrict access to materials for U.S. users.

The specific threats that risky import dependence presents include: restrictions on the physical availability of US raw-material inputs that US manufacturers use, unexpectedly high prices for the inputs when supplies are restricted, and threats to national defense when lack of access to a raw material impairs US military and essential civilian preparedness.

Question 2: How many universities are teaching mining and economic geology in the United States? About how many students each year receive degrees? As the professors who teach those programs retire, will there be individuals to replace them? What does that suggest for the future of mining in the United States?

About 17 universities in the United States teach mining, mineral processing or economic geology. About half of these programs are small and arguably operating below critical mass with five or fewer faculty members. The approximate number of students receiving B.S. degrees annually is 350; M.S. or M.Eng. degrees, 100; and Ph.D. degrees, 45. Many of these graduates are international students. These university programs face significant challenges in recruiting new faculty members given the small number of new Ph.D. recipients each year and the competition from the private sector for their talents. The Society of Mining, Metallurgy and Exploration can provide additional insights into these issues (www.smenet.org).

Question 3: Concurrent with a rapidly aging mining engineering workforce, undergraduate programs in mining engineering are shrinking or closing at an alarming rate. Outside of providing more funding for graduate students, what can be done to grow our university programs and incentivize more young people to go into the mining field?

I support funding for early-stage (pre-commercial) scientific research that will drive innovation in economic geology, mining engineering, mineral processing and extractive metallurgy. This funding should encourage, even require, researchers in these mineral disciplines to reach out and engage researchers in related disciplines such as geochemistry and mechanical, industrial and electrical engineering.

United States Senate Committee on Energy and Natural Resources
March 28, 2017 Hearing: The United States' Increasing Dependence on Foreign Sources of Minerals and Opportunities to Rebuild and Improve the Supply Chain in the United States
Questions for the Record Submitted to Dr. Roderick Eggert

Question 4: Others are paying attention to this issue, especially China, but the United States does not seem to be concerned with our increasing dependence on foreign mineral sources. What do we need to do to make these issues relevant to U.S. policymakers? And, how can we convince companies at the end of the supply chain to start talking about our foreign dependence and the significant problems that it creates?

To make these issues relevant to U.S. policy makers, emphasize:
- *The significant potential for domestic production of mineral raw materials from unconventional sources that can be unlocked through innovation (as happened with oil and gas resources).*
- *The links between (a) mineral raw materials and (b) U.S. manufacturing and consumers of final products. Emphasize the final products that depend on mineral raw materials.*

To convince companies at the end of the supply chain, continue and enhance the activities of the U.S. Geological Survey's National Minerals Information Center, which gathers information and conducts strategic analysis on mineral resources and material flows. My experience with the rare-earth scare of 2010 and 2011 was that many downstream users of raw materials (manufacturers) were almost completely unaware of their vulnerabilities because they were many steps removed from mining and mineral processing of their raw materials; they purchased components and systems rather than mineral ores or concentrates.

This is not to say that the federal government has primary responsibility for managing manufacturers' immediate supply chain risks; it does not. But the federal government plays an essential role in creating the technical, human and intellectual infrastructure that over the longer term responds to short-term supply risks and longer-term concerns about raw-material availability.

Questions from Senator Al Franken

Question 1: In your testimony you highlight the importance of innovation as part of the solution to our reliance on foreign minerals. Essentially, if we are able to improve production processes, waste less, and use less we can cut our overall need. You also underscore the importance of the federal role here in terms of investing in research and development as well as assisting with commercialization through public-private partnerships. I completely agree with you about the importance about federal investment in R&D.

In Minnesota, we have seen the benefits of these types of R&D investments. It was a University of Minnesota professor toiling in obscurity for 33 years who figured out how to make taconite a viable source of iron for steelmaking, at just the time when northern Minnesota was running out of high-grade iron ore.

United States Senate Committee on Energy and Natural Resources
March 28, 2017 Hearing: The United States' Increasing Dependence on Foreign Sources of Minerals and Opportunities to Rebuild and Improve the Supply Chain in the United States
Questions for the Record Submitted to Dr. Roderick Eggert

More recently, University of Minnesota researchers are working to replace Rare Earth minerals in high-powered, permanent magnets with abundant resources like iron. This research has been funded through an ARPA-E grant. The same ARPA-E program that President Trump proposed eliminating in his 2018 budget. I am concerned that cutting programs like this would erode our global competitiveness. You serve as the deputy director of the Critical Materials Institute, which is an innovation hub funded through the Department of Energy. Can you discuss some the breakthroughs that the hub is working on, and how the work would be impacted if the cuts called for in the President's Budget Blueprint are enacted?

The Critical Materials Institute (CMI) is a consortium of Department of Energy national laboratories, universities and companies. CMI aims to develop technological solutions for alleviating supply-chain risks in clean energy technologies such as electric vehicles, advanced lighting, wind turbines and solar power. It conducts early-stage research with an eye toward handing off innovations to the private sector for development. Completing its fourth year of operation, CMI has issued some 50 invention disclosures, 15 patent applications, two technology licenses, two open-source software packages and more than 80 refereed publications. Industrial collaborators are working to incorporate these accomplishments into their products and processes.

Over the next several years, CMI has a number of goals, including:

- *Demonstrating the production of neodymium-iron-boron (Nd-Fe-B) magnets, essential in high-efficiency motors and many military applications, using materials and technologies located entirely within the United States.*
- *Developing a new permanent magnet material that rivals Nd-Fe-B, using reliably available elements.*
- *Developing a new permanent-magnet motor design with optimized system performance, based on printable magnets (that is, additive manufacturing or 3-D printing).*
- *Discovering new red and green phosphor candidates suitable for use in LED lamps.*
- *Demonstrating hard disk drive disassembly rates exceeding 5,000 per day, to enable the recovery of voice-coil motor magnets for recycling or reuse.*
- *Scaling up the supercritical fluid process for dissolution, separation of dissolved components, and refinement of separated critical elements, from milligram to kilogram quantities.*

This and other work would be at risk if budgets are reduced.

Question 2: Are other countries working on similar innovations, and if our federal research and development budget is slashed, are we at risk of ceding ground?

A number of other governments are funding innovative science and engineering that aims to enhance and diversify primary production of mineral resources, improve manufacturing

**United States Senate Committee on Energy and Natural Resources
March 28, 2017 Hearing: The United States' Increasing Dependence on Foreign Sources of
Minerals and Opportunities to Rebuild and Improve the Supply Chain in the United States
Questions for the Record Submitted to Dr. Roderick Eggert**

efficiency and recycling and develop new materials. In particular, China, Japan, South Korea, and the European Union are funding these types of research.

Yes, if our federal research and development budget is slashed, we risk ceding ground.

Questions from Senator Joe Manchin III

Questions: West Virginia University participates in the U.S. Department of Energy National Energy Technology Laboratory's ongoing program to recover rare earth elements from coal and coal byproducts. Acid mine drainage from pre-law coal mines is a source of pollution in the Appalachian region. As it turns out, when these areas are treated to meet current regulatory requirements, the process produces solids enriched in critical rare earth elements. Right now, West Virginia University and its partners work to sample and analyze acid mine drainage solids from 120 acid mine drainage treatment sites at coal mines across the northern and central Appalachian coal basins in West Virginia, Pennsylvania and Ohio. They work together to develop a cost-effective process to treat and recover rare earth elements from the sludges produced at these mines.

Do you agree that this technology would help enhance U.S. supply of critical minerals, thereby better protecting us against supply disruptions?

Yes, this technology has the potential to enhance U.S. supply of rare earths.

Do you see potential here for commercialization of this process?

Yes, recognizing that coal and coal byproducts are one of several unconventional sources for rare earths that, in effect, are competing with one another to become commercially viable. Others include: base metal deposits, such as the Bingham Canyon mine in Utah; phosphate rock and fertilizer production streams in Florida and Idaho; and rare-earth mineralogies for which there are no proven methods of efficient extraction and recovery of rare earths.

Questions from Senator John Hoeven

Questions: In your testimony, you discussed the importance of education for the future of the mining industry. A report that you mentioned in your testimony, you discussed how there is a significant gap between the number of individuals who are being trained to be in the mining profession, and the number of people that we will need in the coming years.

In my state of North Dakota, we are educating and training the next generation of scientists, mathematicians, and engineers to be leaders in the mining industry.

United States Senate Committee on Energy and Natural Resources
March 28, 2017 Hearing: The United States' Increasing Dependence on Foreign Sources of Minerals and Opportunities to Rebuild and Improve the Supply Chain in the United States
Questions for the Record Submitted to Dr. Roderick Eggert

- In order to equip the next generation with the skills necessary to succeed in the mining profession, what are some things policymakers should consider to support the states in educating at the students at the high school and post-secondary levels?
- What policies should Congress consider in order to draw more young people into this field?

The federal government should continue to work with states to emphasize and enhance high-school and post-secondary education in science, technology, engineering and mathematics. I support funding for early-stage (pre-commercial) scientific research and graduate education that will drive innovation in economic geology, mining engineering, mineral processing and extractive metallurgy. This funding should encourage, even require, researchers in these mineral disciplines to reach out and engage researchers in related disciplines such as geochemistry and mechanical, industrial and electrical engineering.

Questions from Senator Mazie K. Hirono

Question 1: In your testimony, you describe the essential role government plays in fostering innovation through research and education. This knowledge is applied to managing supply-chains for materials that are used in clean-energy technologies such as high-efficiency motors, batteries, advanced lighting, and solar materials. How does the research undertaken at the Department of Energy and the Ames Lab's Critical Materials Institute meet a research need that is not otherwise met by the private sector? In your view, how important is early-stage public investment to successfully commercializing innovative energy technologies?

Early-stage research undertaken at the Department of Energy and the Ames Lab's Critical Materials Institute does meet a research need that is not otherwise met by the private sector.

Private companies and individuals certainly have incentives to, and do, invest in research and education because of the benefits they bring to companies and individuals. But from society's perspective, private companies and individuals by themselves underinvest in research and education because the benefits are uncertain, often far in the future and often difficult for companies and individuals to fully capture.

Early-stage public investment is very important to successfully commercializing innovative energy technologies.

Question 2: President Trump's fiscal year 2018 budget proposal included drastic cuts for the Department of Energy, especially for the energy efficiency and renewable energy program. The President's budget proposal calls for cutting $2 billion, or nearly 53 percent, from the Office of Energy Efficiency and Renewable Energy, the Office of Electricity Delivery and Energy Reliability, the Fossil Energy Research and Development program, among other programs. These drastic funding cuts will likely harm the State of Hawaii's efforts to accelerate the

United States Senate Committee on Energy and Natural Resources
March 28, 2017 Hearing: The United States' Increasing Dependence on Foreign Sources of Minerals and Opportunities to Rebuild and Improve the Supply Chain in the United States
Questions for the Record Submitted to Dr. Roderick Eggert

deployment of renewable energy technology in order to meet our goal of achieving 100 percent renewable energy for the electricity sector by 2045. If these funding cuts are enacted, what would be the impact to Ames Lab's CMI research effort?

If enacted, these budget cuts might negatively impact the Ames Lab's CMI research, although the proposals are not sufficiently detailed to know for sure.

Questions from Senator Catherine Cortez Masto

Question 1: You mention that a role for the federal government would be in facilitating the commercialization of ideas created in early-stage R&D through public-private partnerships. Can you comment on successful models in which government improved communication between researchers and commercial developers?

Successful communication occurs when researchers in private, profit-driven organizations help frame, shape and evaluate the research activities of national lab and university researchers.

Successful models are those that create systems for both formal and informal communication between and among collaborators that are in different organizations, in different locations and with different cultures. Formal systems include periodic in-person meetings and reviews that require travel. Informal systems include more-frequent interactions through webinars and virtual meetings that do not require travel but encourage and force interactions.

Question 2: You mentioned that the Critical Mineral Institute is developing technologies for recycling as part of diversifying the supply chain. Could newer technologies motivate companies to invest more in recycling?

Yes. A major barrier to increased recycling is technological; many existing systems are energy intensive and do not recover many potentially recoverable materials. What is needed is more-efficient systems of physically and chemically separating different elements and materials from one another in heterogeneous recycling streams. A major, non-technological barrier to increased recycling of end-of-life products is the expense of collecting and sorting goods once they reach the ends of their useful lives.

Question 3: Mining companies, like Barrick in Nevada, are prioritizing moving toward digitizing their operations to improve communication, efficiency, and worker and environmental safety. Are technologies also being developed to improve overall operations of mining companies?

Mining companies around the world are working, often with suppliers, to incorporate state-of-the-art communications, sensing, automation, robotic and other technological systems into mining operations to improve overall efficiency, safety and environmental performance.

**Statement Submitted on Behalf of
The Minerals Science and Information Coalition**
Submitted by Mark Ellis, Chair

**To the United States Senate
Energy and Natural Resources Committee
Regarding, the Committee hearing to examine the United States' increasing dependence
on foreign sources of minerals and opportunities to rebuild and improve the supply
chain in the United States**
April 11, 2017

Thank you for the opportunity to submit a written statement for the record on behalf of the undersigned members of the Minerals Science and Information Coalition on the Committee hearing to examine the United States' increasing dependence on foreign sources of minerals, and opportunities to rebuild and improve the supply chain in the United States, held on March 28, 2017.

The **Minerals Science and Information Coalition** (MSIC or the Coalition) is a broad-based alliance of minerals and materials interests groups united in advocating for reinvigorated minerals science and information functions in the federal government. Our group is comprised of trade associations, scientific and professional societies, groups representing the extractive industries, processors, manufacturers, other mineral and material supply-chain users, and other consumers of federal minerals science and information. This testimony focuses largely on the testimony provided by Dr. Murray Hitzman, Associate Director – Energy and Minerals, U.S. Geological Survey U.S. Department of the Interior (USGS), and the important role USGS plays in providing and maintaining the minerals science and research that supports every sector of our economy.

Minerals and mineral materials are part of virtually all the products we use every day, acting as the raw materials for manufacturing processes or as the end products themselves. Minerals are contained in buildings, roads and civic infrastructure projects. They also are used in the manufacture of paper, glass, ceramics, plastics, refined metals, and a host of intermediary materials. These, in turn, find their way into the manufactured products that make up our daily lives: automobiles, mobile phones, and computers. They are critical ingredients in specialized applications for national defense and energy technologies. The mining industry underpins the high standard of living we enjoy and to which we've grown accustomed. Every sector of industry relies on a variety of minerals to generate their end products, making a stable and reliable supply chain critical for the continued growth and success of our economy.

Given the vital national importance of minerals science and information, MSIC commends you, Chairwoman Murkowski, for recognizing the need for greater understanding in minerals science and information, and the global mineral supply chain. The Coalition applauds the Committee for calling this hearing to discuss the United States' growing dependence on mineral imports, and would like to take the opportunity to offer support for some measures discussed during the hearing.

MSIC generally supports increased investment in minerals sciences, specifically through federal funding for the important work of USGS. As testified to by Dr. Hitzman, USGS provides important information on the "current production and consumption for 84 mineral commodities, both domestically and internationally for 180 countries" in the form of the annual *Mineral Commodities Summaries*. USGS is one of the only public and unbiased sources of this information that is used by industry and governments around the world. USGS's National Minerals Information Center (NMIC), in partnership with the Department of Energy developed "an early warning screening tool to identify critical minerals of concern for economic and national security and stay ahead of the curve as technology changes and geopolitical unrest shifts," which is another example of the invaluable work of USGS. This program, in particular, highlights the importance of investment in forecasting tools as a way to safeguard the supply chain. Additionally, MSIC was pleased to note the Chairwoman's interest in expanding USGS's geological mapping capabilities through federal investment. The USGS National Cooperative Geological Mapping Program is the primary source of funds for geological mapping in the U.S., and acts to foster partnerships at the Federal, State, and university levels.

Finally, MSIC offers its support for the reintroduction of S. 883, the American Mineral Security Act, in the 115th Congress. As in years past, the Coalition believes strongly in the stated goals of S. 883 to strengthen and improve our understanding of critical minerals and to develop a robust scientific and statistical information and forecasting capability to identify and anticipate threats to supply chains. The recent crisis in the global supply of rare earth elements caused by Chinese export restrictions is a case study in the importance of a stable mineral supply chain. Supply chains can be long, complex, and vulnerable to disruption for many reasons. The restrictions in the supply of rare earths to the U.S. threatened the production of components that are essential for U.S. defense systems, in addition to a vast array of communications, clean energy, electronics, automotive, and medical products. Both the private and the public sectors realize that we must reduce risks to our supply chains. But we cannot do this without accurate, timely information on the nature, location, and characteristics of our domestic mineral resources, and on the worldwide supply of, demand for, and flow of minerals and materials. This information is the foundation for identifying and forecasting existing and emerging vulnerabilities, and for sound decision making by business leaders and policy makers.

The USGS plays a vital role in allowing leaders in our businesses and governmental institutions to make decisions based on the best information available on our resources. It is the Minerals Science and Information Coalition's belief that prioritizing both the science and information components of USGS's Mineral Resources Program and the National Minerals Information Center is vitally important to our national defense and economic well-being. As such, the Coalition applauds the recent hearing for raising awareness of our national mineral resources and the importance in federal reinvestment in our nation's ability to continue to develop and grow responsibly by using our own resources.

Thank you for the opportunity to submit a statement for the record of this hearing.

    American Chemical Society

    American Exploration and Mining Association

    American Geosciences Institute

    American Physical Society

    Industrial Minerals Association – North America

    International Diatomite Producers Association

Mining & Metallurgical Society of America

National Industrial Sand Association

National Mining Association

National Stone, Sand and Gravel Association

Contact information: Ariel Hill-Davis, Director, Industry Affairs, Industrial Minerals Association – North America, 1200 18th St NW, Suite 1150, Washington, D.C. 20036. arielhilldavis@ima-na.org. 202-457-0200.

www.ingramcontent.com/pod-product-compliance
Lightning Source LLC
Chambersburg PA
CBHW062223220526
45471CB00009B/3320